MAKING
LEARNING
TEACHING
EXPLORING

FAMILY
Maker Camp

www.makercamp.com

CONTENTS

28

COLUMNS

FEATURES

ON THE COVER:
Physiotherapist Marine Sacco wears a snorkel mask modified with a 3D printed adapter to fit an air filter, while working in a Covid unit of a Belgian hospital.
Credits: Albert De Beir–VUB/Brubotics, Adobe Stock–korkeng (background), and Mike Senese (cell phone).

COMBATING COVID-19

56

70

78

84

96

Adobe Stock – Feydzhet Shabanov and alexxndr, MakerMask.org, Joe Hoddinott (dphojoegraphy, Keith Hammond, Taylor Tabb, Katie Rosa Marchese, Ulrich Schmerold, Ekaggrat Singh Kalsi

Make:

> "[Makers are] making whatever we can get them to make."
> — Jay Margalus, Illinois PPE Network

PRESIDENT
Dale Dougherty
dale@make.co

VP, PARTNERSHIPS
Todd Sotkiewicz
todd@make.co

EDITORIAL

EXECUTIVE EDITOR
Mike Senese
mike@make.co

SENIOR EDITORS
Keith Hammond
keith@make.co

Caleb Kraft
caleb@make.co

PRODUCTION MANAGER
Craig Couden

CONTRIBUTING EDITOR
William Gurstelle

CONTRIBUTING WRITERS
Adnan Aga, Heinz Behling, Tim Deagan, Julie A. Finn, Graffiti Research Lab, Saul Griffith, Slater Harrison, Justine Haupt, Ekaggrat Singh Kalsi, Bob Knetzger, Mario Marchese, Jorvon "Odd Jayy" Moss, Samer Najia, Niq Oltman, Danny Osterweil, Charles Platt, Ulrich Schmerold, Douglas Stith, Tim Sway, Taylor Tabb, Cy Tymony, Matthew Wettergreen

DESIGN & PHOTOGRAPHY

CREATIVE DIRECTOR
Juliann Brown

CONTRIBUTING ARTISTS
Alison Kendall, Charles Platt, Kirk von Rohr, Julie West

MAKE.CO

ENGINEERING MANAGER
Alicia Williams

WEB APPLICATION DEVELOPER
Rio Roth-Barreiro

GLOBAL MAKER FAIRE

MANAGING DIRECTOR, GLOBAL MAKER FAIRE
Katie D. Kunde

MAKER RELATIONS
Sianna Alcorn

GLOBAL LICENSING
Jennifer Blakeslee

BOOKS

BOOKS EDITOR
Patrick DiJusto

MARKETING

DIRECTOR OF MARKETING
Gillian Mutti

OPERATIONS

ADMINISTRATIVE MANAGER
Cathy Shanahan

ACCOUNTING MANAGER
Kelly Marshall

OPERATIONS MANAGER & MAKER SHED
Rob Bullington

PUBLISHED BY

MAKE COMMUNITY, LLC
Dale Dougherty

Copyright © 2020 Make Community, LLC. All rights reserved. Reproduction without permission is prohibited. Printed in the USA by Schumann Printers, Inc.

Comments may be sent to:
editor@makezine.com

Visit us online:
make.co

Follow us:
🐦 @make @makerfaire @makershed
📘 makemagazine
📷 makemagazine
▶ makemagazine
📺 twitch.tv/make
Ⓟ makemagazine

Manage your account online, including change of address:
makezine.com/account
866-289-8847 toll-free
in U.S. and Canada
818-487-2037,
5 a.m.–5 p.m., PST
cs@readerservices.makezine.com

Make:
Community

Support for the publication of *Make:* magazine is made possible in part by the members of Make: Community. Join us at make.co.

CONTRIBUTORS

What "shelter at home" tip has been most effective for you?

Justine Haupt
Mattituck, NY
(Digital Revolution)
Embracing the opportunity to work on projects I wouldn't otherwise have had time for.

Jorvon Moss
Compton, CA
(My Monkey Companion Bot)
Keep busy, work on a big project you can share it with the world when the pandemic is over.

Samer Najia
Alexandria, VA
(Pick and Pull PC Projects)
Stay in shape and indulge your hobbies. Every so often open a window and let in some fresh air/sunshine.

Issue No. 73, Summer 2020. *Make:* (ISSN 1556-2336) is published quarterly by Make Community, LLC, in the months of February, May, Aug, and Nov. Make Community is located at 150 Todd Road, Suite 200, Santa Rosa, CA 95407. SUBSCRIPTIONS: Send all subscription requests to *Make:*, P.O. Box 17046, North Hollywood, CA 91615-9588 or subscribe online at makezine.com/offer or via phone at (866) 289-8847 (U.S. and Canada); all other countries call (818) 487-2037. Subscriptions are available for $34.99 for 1 year (4 issues) in the United States; in Canada: $43.99 USD; all other countries: $49.99 USD. Periodicals Postage Paid at San Francisco, CA, and at additional mailing offices. POSTMASTER: Send address changes to *Make:*, P.O. Box 17046, North Hollywood, CA 91615-9588. Canada Post Publications Mail Agreement Number 41129568. CANADA POSTMASTER: Send address changes to: Make Community, PO Box 456, Niagara Falls, ON L2E 6V2

PRINTED WITH
SOY INK

Make: Projects

Join the community of over 25,000 sharing 8,000 plus projects online!

Make: Projects is a collaborative platform that significantly improves the way to document, share and advance projects online. It brings together makers, engineers, inventors, and doers to develop passions or solve our society's most persistent challenges. The messy process of making is at the heart of the platform! It's both a private space to tinker and organize ideas with a few friends, paired with a public forum to showcase your work and get feedback.

www.makeprojects.com

Completing the collection and building your first pizza oven

Will Kinney

PIZZA TIME!

Thanks for publishing this great design! ("One-Day Wood-Fired Pizza Oven," *Make:* Volume 53, page 34) I looked around at a lot of build plans, and most of them were dauntingly complex, and seemed to require a lot of expertise with brick and mortar to pull off. Your design was easy for a beginner to get right. Really nice.

—*Will Kinney, Buffalo, NY*

Prison Ban of the Month

- » **LOCATION:** Arizona Dept. of Corrections, Rehabilitation & Reentry
- » **TITLE:** *Make:* Vol. 72
- » **REASON:** Depictions, descriptions, and instructions regarding the function of locks and/or security devices (e.g, cameras, alarms) or how to bypass or defeat the security functions of these devices.

OPR - ADC <OPR@azadc.gov>   Fri, Mar 6, 2020 at 2:11 PM
To: editor@make.co

To Whom It May Concern:

The Arizona Department of Corrections, Rehabilitation & Reentry has determined that your publication described below contains Unauthorized Content as defined in Department Order 914.07 and, as a result, may be released in part or excluded in whole for the specific reason(s) given below.

Publication Title: Make: Magazine

ISBN: **Volume/Number:**   72   **Publication Date:**   Spring 2020

Reason: DO 914 -7.2.5 Depictions, descriptions and instructions regarding the function of locks and/or security devices (e.g, cameras, alarms) or how to bypass or defeat the security functions of these devices.

You and/or the inmate subscriber may appeal the decision by notifying us via email or U.S. Mail within 30 calendar days after you receive this notice. By appealing, you consent to allowing OPR to redact any Unauthorized Content within the parameters set forth in Department Order 914.06 § 6.13. Your consent is strictly limited to authorizing ADCRR to alter by redaction your publication. It does not constitute consent to the substance of the actual redaction(s) subject to this Notice.

By email to   Include "Request Appeal of" and the publication title in the subject line.

By U.S. Mail to   OPR
                       Arizona Department of Corrections, Rehabilitation & Reentry
                       1601 West Jefferson Street
                       Mail Code 481
                       Phoenix, AZ 85007

The Department will notify you of the final decision within 60 days of receiving your request. The appeal decision is final.

Sincerely,

Arizona Department of Corrections, Rehabilitation & Reentry
Office of Publication Review

Tasty Tweets:

Eliza McCarthy
@ElizaMcCarthy7

We were given a subscription to Make magazine, and idly turning pages, find long feature by Saul Griffiths, a very charming climate hero, on how to decarbonize!! Highly recommend: makershed.com/collections/ma...

1:07 PM · Mar 8, 2020 · Twitter Web App

1 Retweet   **3** Likes

Tested ✓
@testedcom

Behold a COMPLETE SET of @make magazines. Care to share photos of one of YOUR complete collections of something?

12:00 PM · Apr 4, 2020 · Buffer

12 Retweets   **214** Likes

Victoria McGovern
@BWFPATH

The second weekend of working from home, the back issues of Make magazine took on a curious allure.

9:50 AM · Apr 11, 2020 · Twitter for iPhone

Allie Weber
@RobotMakerGirl

3rd grade me in Make Magazine. 😄

Tested ✓ @testedcom · Apr 4
Replying to @RobotMakerGirl and @make
😊

4:39 PM · Apr 4, 2020 · Twitter for iPhone

2 Retweets   **48** Likes

MAKE: AMENDS

Reader Richard Spenser wrote in with notes and suggestions from his build of the "**Best-Yet DIY Coffee Roaster**" (*Make:* Volume 71, page 42). In response, author Larry Cotton has added these corrections. Thanks Richard!

1. In Figure **A**, for the large base, change 9"×20" to 9"×22"

2. In Figure **B**, for the sifter pivot support, remove the 11¾" dimension. Then add a ½" dimension from the rounded top of the support down to the 5⁄32" hole. (Yes, I know it's not equivalent, but it's more accurate and easier to locate.)

3. In Figure **D**, for the crank rotator, change the hole-to-hole dimension from 15mm (.591") to 16mm (.630").

Everything's Changed

by Mike Senese, Executive Editor

Adobe Stock – astrosystem

It goes without saying, but what a start to a year, for us and for everyone.

We began laying out the framework for this issue last December. At the beginning of January we shipped our previous issue the printer and shifted focus to generating this new edition — assigning and approving articles, traveling to events, conducting interviews, and writing pieces. Normal magazine production processes.

It feels odd to now look back at those months, knowing that what was developing elsewhere would turn into an unprecedented worldwide shutdown.

We pushed forward with the issue into the start of the spring, but as Covid-19 made its way from country to country, closing schools, businesses, and entire communities, and as crucial medical supplies suddenly became impossible to find worldwide, we shifted our efforts to online coverage of how the maker community was stepping up — using their tools and training to combat the effects of the coronavirus (read those stories at makezine.com/go/plan-c). We've never seen a globally distributed force rapidly coalesce like this, determining needed equipment, then designing, producing, and delivering it to front-line medical providers, while navigating material shortages and implementing necessary modifications in real time.

With a few weeks left in the magazine production cycle, it became apparent that this was much bigger than the themes we were working on for the issue. So we started over, generating an entirely new special section based on the people and projects that have been helping in the fight against this virus, and the lifestyle changes we're now navigating.

What you're seeing in these pages is a snapshot of the Covid-19 response at mid-April 2020. When this issue of *Make:* gets to your house, it'll likely be outdated. But the stories, the tools, the output, they're all part of a new world.

Our immense appreciation goes out to everyone that's working tirelessly and selflessly to keep us going — the medical professionals, the makers, the volunteers, the delivery people and food workers and everyone else. We hope that it's over when you read this letter. It doesn't seem like it will be. But remember, there's never been a challenge the human spirit couldn't overcome, and this will prove to be the same. With everone's help, we can do this. ◗

MADE ON EARTH

Backyard builds from around the globe

Know a project that would be perfect for Made on Earth? Let us know: *editor@makezine.com*

Joe Hoddinott (@phojoegraphy)

THE SOFTER SIDE OF METAL ELLENDURKAN.COM

Delicate. Soulful. Evocative. These aren't typically the terms you hear us throw around when we speak of a blacksmith's work. However, this artist is anything but typical.

Ellen Durkan, known as **Iron Maiden Forge**, has been creating elaborate and mesmerizing works of fashion that are born from hot steel on the anvil. At a glance you might think a piece is a soft neck ruffle, accentuating a flowing gown, but upon further inspection you'll find that the curves are frozen in time, molded in steel or copper by a skilled hand.

While she was a student in art school in Delaware, Durkan worked with steel frames, draped with fabric. She found herself drawn to the metal frames themselves, ultimately driving her to begin teaching herself forging. All of her art starts with the human form and in her words,

needs the human form to be properly displayed.

"The human figure has always been an important part," she says. "I think my work, in order to be experienced properly, needs to be on the body. Metalwork is just part of the art."

The shapes that evoke the curves of the human body can morph to living plants or weave into intricate leather works of fashion. Being self taught, Durkan's work is ever evolving. Her most recent area of expansion is repoussé, allowing her to make even more finely sculptured structures that can mimic the human form — not only a figurative layer to her work, but in some cases even a literal layer of art. See her piece *Speak* (above and left), which involves a feminine face, trapped behind a finely formed mask with an impassive expression, trapped again behind a wrought iron cage. —*Caleb Kraft*

makeprojects.com 11

PULP FICTION

INSTAGRAM.COM/PAPER_CARRIZALES

The term "paper mâché" brings up fond memories of youthful summer camp projects — strips of newspaper dipped in paste, layered on top of a balloon, face, or cardboard model, then left to harden and be painted. Those resulting oblong balls, lumpy masks, and indeterminate piñatas should be familiar to us all.

The centuries-old medium isn't well known for its finer art aspects, but there are some creators out there making pieces of respectable quality, often geared toward youngsters. And then there's Jalisco, Mexico-based **Fernando Pérez Carrizales**, who uses paper mâché to create studio-quality renditions of fantasy and sci-fi creatures and elements.

His pieces range from small and detailed to immense and intimidating; a full-scale Pale Man from Guillermo Del Toro's *Pan's Labyrinth* and a wall-sized face of a Rancor jump out in particular. There are a lot of familiar images and mashups; his inspiration, however, comes from history and heritage. "We all grew up with monsters around us in cartoons, movies, books," Carrizales says. "The stories our grandmothers told us always include fantastic beings. And in the case of Mexico's pre-Hispanic culture, it's full of monsters and warrior eagles and jaguars. If we add Día De Los Muertos, it's hard not to incorporate fantasy into sculptures."

Carrizales followed his brother in making art, but stepped away from it for a period until a tragedy brought him back. "My daughter died when I was 21, so I took up sculpture as a form of therapy," he says. "It was my refuge!" He and his brother are now giving art classes and workshops to local kids and parents alike.

"I have used materials like modeling clay, wax, epoxy, resins, foam rubber — and I always end up going back to paper," Carrizales says. "It can be carved like wood, shaped like clay; it can be applied in casting molds, allowing you to combine many techniques. Furthermore you can find it anywhere, often discarded, making it an underestimated raw material." —*Mike Senese*

Fernando Carrizales, Mike Senese

SMALL-SCALE SPACE TWITTER.COM/ROKUBUNNNOICHI

The video zooms in on what looks like a standard power outlet wall plate. Then a hand reaches into frame and opens the outlet like a tiny doorway to reveal a detailed miniature living space replete with illuminated desktop computer, microwave, blinking router lights, mini-mini fridge, and everything else you'd expect to find in an efficient workspace, just at a microscopic size.

Laboring over three months, artist **Mozu** (aka Mizukoshi Kiyotaka) used plastic boards, LED lights, wood, and more to make his *Miniature Secret Base Inside an Outlet*. The most difficult part, he says, was making the blinking LEDs for the Wi-Fi router. Living in Tokyo, Japan, Mozu has been creating miniature models since he was 16 years old. It's an ongoing childhood dream of his,

adorable stop-motion animation *Rilakkuma and Kaoru* on Netflix, and recently started his own design studio (mozustudios.com).

The magic of his tiny, hidden office isn't just that the scene is so small, it's how much the space looks lived in. The RCA plugs from the video game system are connected to the front of the TV because plugging into the back is a pain; the slight curl of the calendar on the wall suggests a person who regularly looks ahead at upcoming events.

At one point I had forgotten that I had the video open in a browser tab, and when I clicked back I legitimately thought another video of a *real* office had started autoplaying. But nope — it was still Mozu's tiny world. It's that good. —*Craig Couden*

WAYS TO READ Make:

IN PRINT

ON THE GO

ONLINE

New Power Nation

Written by Dr. Saul Griffith

HOW WE'LL REMAKE AMERICA'S NATIONAL INFRASTRUCTURE TO ELECTRIFY EVERYTHING

DR. SAUL GRIFFITH is founder and principal scientist at Otherlab, an independent R&D lab, where he focuses on engineering solutions for a clean energy, net-zero carbon economy. Occasionally making some pretty cool robots too. Saul got his PhD from MIT, and is a founder or co-founder of makanipower.com, sunfolding.com, voluteinc.com, treau.cool, departmentof.energy, materialcomforts.com, howtoons.com, and more. Saul was named a MacArthur Fellow in 2007.

I'm a scientist, engineer, inventor, and father who is passionate about my kids being able to live in a clean world and feel the sense of awe in nature that I've been lucky to enjoy. I'm in this fight with all I've got, including a lot of data that convinces me that it's rational to have hope — that we can win big against this climate emergency.

And if we win — *when* we win, because there is no other option — we'll be much better off than before. When we replace fossil fuels with clean electricity, we'll not only have a better future for our kids, we'll create new jobs and remain the economic powerhouse that we are today.

Billionaires might dream of escaping to Mars. The rest of us, frankly, have to stay and fight.

It's a climate emergency. Break the glass.

What's the Clean Energy Infrastructure We Need?

In short, to electrify everything, we'll need about three times the amount of electricity that we currently produce.

Today, the U.S. grid delivers 450GW (gigawatts) of electricity. If we electrify nearly everything, we'll need about 1500GW, or 1.5TW (terawatts). That's a lot. That means on this path to decarbonization we'll need more than 3 times as much electricity. How do we get there?

Today we can produce electricity at remarkably low costs, but the costs of distributing that electricity remain high. In the U.S., the average cost of grid electricity is 13 cents per kilowatt-hour (¢/kWh). Amazingly, *more than half* of this is the cost of transmission and distribution:

7.8¢/kWh. In contrast, rooftop solar in Australia provides electricity to the customer at just 6–7¢/kWh total.

That should seem shocking, and let's reflect upon it for a moment. Locally generated electricity, because it nearly eliminates transmission and distribution costs, will likely always be less expensive than any centrally generated power source. The cheapest energy in the future will likely come from your solar roof, and we should generate as much of it as possible. In addition to homes we should look to the roofs of commercial buildings and solar cells over parking lots to increase the local generation capacity and keep our costs as low as possible.

The total amount of electricity we need, however, is more than can fit on all of our rooftops and parking spaces. So we'll also need a significantly expanded electrical grid to supplement local generation with electricity generated in large centralized facilities. It will supply electricity to places that need it from other places where the sun is shining or the wind is blowing or the reactors are reacting. It will have microgrids and household grids and neighborhood grids and a giant grid to connect them all together. This "energy internet" will keep transmission and distribution costs as low as possible while balancing supply and demand.

The exact details will vary geographically: by local population density (urban vs. suburban vs. rural), by climatic region (hot vs. cold vs. temperate), and by resource availability (sunny vs. windy vs. soggy places that can generate hydroelectricity). Places with lower density, mild climates, and good solar resources (like Australia, California, New Mexico, and Texas) can almost completely solve the challenge with well-managed solar alone.

Multi-story dwellers in New England, however, have too small a roof and too cold a winter for this solution. High-density populations (in any climate) will probably need to lean more heavily on nuclear power or some other imported energy, which could be long-distance electricity transmission, renewably generated hydrogen, or biofuels.

In any case, the fastest, most easily scalable way to zero emissions is through electrification. To solve the climate emergency, we need to get shovels in the ground and wires in our walls as soon as possible. We can no longer afford to wait.

RAMPING IT UP

The total electricity we will need to power our lifestyles is 1.5TW–2TW (1 terawatt = 1 trillion watts). We're currently at 0.45TW. How do we ramp up our infrastructure to provide all our energy needs?

We can generate 0.25–0.75TW of solar on our residential roofs alone. If we cover lots of our parking lots and parking spaces, that's another 0.5TW of solar energy. Our densest cities can't rely on solar in the same way, but covering our abundant suburbs, rural areas, parking lots, and commercial buildings will be a big start.

Midwestern farmers are already seeing economic advantages with the profusion of wind turbines silently making them money while cattle graze or crops grow below. More money will be spent in, and remain in, local communities as we build out this infrastructure, whether it's solar on roofs or wind turbines on farms.

Remember, half the cost of electricity is the grid. The more energy we generate closer to our homes and workplaces, the lower our energy costs will be. Our cars and our home heating systems will be the biggest batteries we have when they're interconnected to our new multitude of microgrids. We'll need to place as much solar as possible where we live to reduce transmission and distribution costs.

Our farms and suburbs will be the cornerstone of our new electric infrastructure for both generation and storage. We'll be building infrastructure that guarantees local jobs long into the future. There are even more jobs in retrofitting everyone's basement with a heat pump and their kitchen with new appliances.

The composition of the future grid will be determined mostly by geography; the rest will be hammered out by policy, the market, and people's preferences. The good news is that the target is more than reasonable with known technologies! In many markets solar is by far the cheapest form of electricity. With high certainty we can say that in a decarbonized future, average U.S. families will pay much less for all their energy bills than they do today.

Yes, it will take energy to create this new 21st century decarbonized energy infrastructure. Solar panels, wind farms, electric cars, and heat pumps don't grow on trees. But the return on investment will be enormous. Reducing the

20th vs. 21st Century Infrastructure

In the 20th century, *public infrastructure* was large and centralized — power plants, transmission lines, bridges, roads, and dams — and much of it was built to support an economy run on fossil fuels:

- 8.8 million lane-miles of roads
- 170,000 gas stations
- 500,000 bridges
- 101 nuclear reactors
- 800,000 miles of sewer lines
- 84,000 dams and reservoirs (not all hydroelectric)
- 200,000 miles of high voltage transmission lines
- 5.5 million miles of local distribution lines
- 4.4 million miles of gas pipelines
- 72,000 miles of oil pipelines
- 130+ oil refineries

Our 21st infrastructure will include more *personal infrastructure* connected to the public infrastructure — climate fixes we can engage in every day (see *Make:* Volume 72, "Decarbonization Begins at Home," makezine.com/go/fop2) — almost all of which help us generate or store clean energy:

- 120 million solar roofs
- 253 million cars
- 16 million trucks
- 90 million furnaces
- 150 million refrigerators
- 100 million hot water heaters
- 1 billion parking spaces
- 120 million new smarter fuseboxes
- Power meters that go both ways

total energy used in our economy by 50% means that these new investments will be profitable. They will create new industries, and they will put millions of people to work.

The main thing standing in our way is how we're going to pay for it. If only the wealthiest people can opt in to paying for climate solutions, we don't have a climate solution. We need to think about finance.

How Do We Pay for It?

You don't fight a war because you can afford it — you fight a war because you can't afford not to. But the fact is, we *can* afford it, and decarbonizing will save all of us a lot of money. How?

First, clean energy is already the cheapest option, and it's getting cheaper every day. Second, we can finance the clean energy future in order to start the transition today. Finally, we can and must pay for our fossil fuel past in order to transition safely to a decarbonized future.

CLEAN ENERGY IS CHEAPEST, AND GETTING CHEAPER

Up until now, clean energy has been developed in places where the economic benefits were obvious. Australia figured out solar because the grid is so distributed that retail grid electricity is expensive. South Australia proved out grid-scale batteries because that was cheaper than new gas plants. California led the world in electric vehicles because air pollution in Los Angeles and other cities made the need clear. Europe and Japan mastered heat pumps because of their cold weather and limited domestic natural gas.

Once adopted, the economic benefits of clean energy are clear. The average electricity price in the U.S. is 13¢/kWh, but rooftop solar hits 6¢/kWh with the right regulations. Right now, if you drive a 25mpg vehicle at $3/gallon, it costs 12¢/mile; a 300Wh/mile electric vehicle using 6¢/kWh electricity will cost only 2¢/mile. At the average natural gas prices in the U.S., heating your home costs $11.20/MBTU. With a heat pump of COP 3 and an electricity price of 6¢/kWh, heat will cost you $5.86/MBTU instead.

And renewable technologies are just getting cheaper thanks to innovations in technology. The solar and wind industries are *learning*, getting

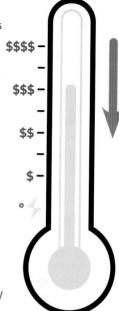

cheaper as production ramps up and innovations overtake the field. These trends can be quantified into a *learning rate*, or the amount the unit cost of a technology is reduced when investment is doubled. For instance, solar is learning at 23% and wind at 12% — as fast or faster than fossil fuels during their 20th-century heyday. Lithium ion batteries are learning at a rate of 17%, dropping from over $1,000/kWh in 2010 to $150/kWh today, with projections to hit $60/kWh by 2030.

Energy will be cheap, and we can save more money by creating a bidirectional grid, allowing energy to flow both to and from the consumer. The fact that our new infrastructure will be closer to home, where more energy is generated and used locally, will make energy even cheaper; our own cars and homes can be used as batteries to store and transmit energy.

So, we've figured out the cheapest sources of clean energy. Why haven't we implemented them? The main reasons are regulations and incentives that favor fossil fuels, and financing.

Regulations are a serious impediment to the market penetration of clean technologies. When you buy solar on your rooftop in Australia it costs $1.20/W, but for reasons of regulations, permitting, and high sales cost, that price is $3/W in the U.S. The underlying hardware is incredibly cheap, with solar modules selling internationally at 35¢/W and believable pathways to 25¢/W. We must update regulations and permitting processes that stand in the way of converting to clean energy.

Likewise, the system of subsidies to support fossil fuels gives these old technologies an unfair advantage. We have to minimize the cost of solar and wind with the right incentives and regulations, while eliminating supports that unfairly discount fossil fuels.

HOW DO WE PAY FOR THE FUTURE?

We've seen that with proper regulation, clean energy is already cheaper than fossil, but financing remains a major barrier to adoption. Solar, wind, electric vehicles, and heat pumps all cost you more up front, but save you money later.

The key to transitioning quickly to renewables will be creating the same kind of public-private partnerships and innovative capital financing strategies that have long underpinned America's economic engine: loans. We must invent the climate loan, a low-interest financing option to help consumers afford the capital investments for 21st century decarbonized infrastructure.

America's lifestyle was built on loans; the car loan and home mortgage were both 20th-century American innovations. The modern mortgage market was shaped by the federal government's intervention in another time of crisis: the Great Depression, when property values plummeted and about 10% of all homeowners faced foreclosure. The government stepped in during Franklin Delano Roosevelt's New Deal, when Congress passed the Home Owners' Loan Act of 1933 to provide low-interest loans for families at risk of default. As a result, hundreds of thousands of homeowners were able to pay their mortgages, and the program actually turned a slight profit. This program gave rise to Fannie Mae in 1938, and then Freddie Mac in 1968, creating the lowest-cost debt pool the world had ever seen.

To win climate stability and a more robust energy infrastructure, the U.S. government must be just as audacious in financing zero carbon capital. Tomorrow's infrastructure will necessarily be more personal and distributed, so it's time to help consumers get the same low interest rates the utilities get. Today, an energy utility can get interest rates below 4% to build yesterday's infrastructure, but a consumer gets stiffed with 9%–12% when they buy solar panels, heat pumps, electric vehicles, and batteries.

As I write this sentence, 3.45% is the U.S. 30-year mortgage rate. If we finance solar panels at this rate, their electricity will cost just 4.5¢/kWh. If, however, the same installation is financed at 10%, as is common today, the same electricity costs 8¢/kWh, nearly twice as much.

If done right, innovative low-cost financing can be one of the most effective ways to ensure equity and universal access to cheap, reliable energy in the 21st century.

HOW DO WE PAY FOR THE PAST?

We must also think carefully about the economic ramifications of the transition away from fossil fuels.

Digging holes in the ground costs money. Finding the one with oil in it costs more money. Fossil fuel companies spend a lot to find fossil fuels, and only recoup those investments slowly over time. This business model requires borrowing money to dig the holes, and when they write that mortgage to the bank, the asset they pledge to the mortgage is the oil coming out of their last well.

In the context of decarbonization, lingering debts like these are called *stranded assets*, and they're a big problem. Stranded assets are resources that once had value but no longer do, usually because of a change in technologies, markets, or social habits — like railroad tracks abandoned due to a shift to automobiles.

Currently, it's estimated that the total debt attributed to fossil fuels that aren't even dug up yet is $135 trillion. Despite the fact that no human has laid eyes on these fossil fuels, they appear as assets on energy companies' ledgers. Climate scientists agree that burning those reserves would compromise the 1.5°C warming limit; indeed, to stay under that target, we must not burn a third of the oil, half of the gas, and 80% of the coal in that asset pool. Because these fuels are already financed, however, they appear as assets on energy companies' ledgers, and they're already traded like any other form of money. If you had $135 trillion in the bank, would you relinquish it without a fight?

A 2018 study in *Nature Climate Change* estimated that as much as $4 trillion would be wiped off the global economy by stranding fossil fuel assets. By comparison, a loss of only $250 billion triggered the crash of 2008. The rippling effects of such an event could be catastrophic.

In navigating this precarious scenario, the best strategy may be to treat the owners of these assets, the fossil fuel industry, as friends rather than enemies. Rather than make deniers

How to Finance World War Zero

» **Minimize the cost of solar and wind** with the right incentives and regulations. This means detailed local regulatory and building code work, as well as reform of National Electrical Codes (NEC) and FERC.

» **Eliminate all fossil fuel subsidies** to even the playing field.

» **Minimize grid costs** by allowing energy to flow both to and from the consumer. This will require utility reform.

» **Finance industrial manufacturing infrastructure** to lower the costs of EVs, batteries, solar, heat pumps, and smart grid components, similar to how the U.S. government underwrote the Arsenal of Democracy.

» **Climate loans** — Create a federal government-guaranteed, low-interest consumer financing instrument similar to Fannie Mae for the package of decarbonizing infrastructure that decarbonizes a home.

» **Buy the proven reserves of fossil fuel companies** to bring them into the green future as allies, not enemies.

and fighters out of these companies, what if we engage them as the best allies to build the decarbonized future? They're extremely good at financing capital-intensive businesses. They have enormous teams of smart and competent people who are good with shovels. Those people could be just as happily — probably more happily — employed building decarbonization infrastructure. Why don't we invite them to be a driving force in World War Zero?

The only roadblock is the stranded assets, so what if we buy them out? They would only ever make a slim profit margin (around 6.5%) anyway. Let's round it up to 10% to be generous: 10% of 135 trillion dollars is $13.5 trillion, a small fraction of our $100 trillion global GDP. The fossil fuel companies would have a huge

amount of capital they could invest in the new energy economy and the new infrastructure of the 21st century, generating jobs and building new businesses with valuations far exceeding that of the stranded assets.

This may sound like a crazy idea, but it's the type of thinking we must embrace to solve our climate problems with the biggest energy infrastructure buildout ever to occur. ⊘

Stand with the Children

The planet is cooking and it's 2020 already — we're going to have to fix this ourselves. This is essay #3 on how to decarbonize our world. Read the series, find hands-on projects to make a difference, and share your ideas at make.co/fix-our-planet.

Native Optimization
Maker Faire on the Navajo Nation
Written by Keith Hammond

Any Maker Faire is a reaffirmation that we're all born to make — it's what makes us human in the first place. It's also a celebration of making's past and future, where traditional crafts meet new technology. And there's nothing like attending a Faire in another country to really fire up those feelings of connection. This spring I was lucky to attend one in a sovereign nation on U.S. soil: Diné Maker Nation Maker Faire, at Navajo Technical University (NTU) on the Navajo reservation in Crownpoint, New Mexico.

The Navajo (or Diné, *the people*, as they call themselves) are globally recognized makers of the American Southwest. They've built a long tradition of fine craftsmanship in wool weaving, silversmithing, and jewelry making, despite bitter headwinds of conquest and deprivation. After centuries of war with the Spanish, Mexicans, and Americans, in 1864 some 8,500 Navajo were force-marched 400 miles to a squalid camp in eastern New Mexico. Hundreds died en route. The Treaty of 1868 finally returned the Diné to their ancestral lands.

The Navajo Nation is the largest Native reservation in the United States, covering an enormous swath of high-desert canyon country on the Colorado Plateau in Arizona, Utah, and New Mexico. But even there, they've often been treated as second-class citizens. Denied water rights to the Colorado River, even as the states around them drank their fill. Poisoned by radiation in the uranium mines and mills that used them as cheap labor without protection. Nuked 100 times by atomic fallout from the Nevada Test Site. Life on the rez was mainly subsistence agriculture and livestock grazing; good-paying skilled jobs were few. Roads were so bad that the school buses often couldn't run; schools were so bad that parents sent kids away to neighboring states.

In the 1970s the Navajo Nation became home to the West's largest coal-fired power plant. It brought a thousand jobs but smogged the reservation — and the Grand Canyon — and was America's third-largest emitter of greenhouse gases. Power went to L.A., Vegas, and Phoenix, while half the reservation still lacked electricity.

That's the past. The last trainload of coal was burned in 2019; today the plant is being

Keith Hammond is senior editor at *Make:* and a former resident of the Colorado Plateau canyon country in Utah. He first visited the Navajo Nation in 1984.

Navajo Technical University amid the mesas of Crownpoint, New Mexico.

dismantled because fracked gas and solar are cheaper. The Navajo are moving forward, getting elected to county commissions, fixing roads, demanding their share of precious water. They're rededicating their schools and colleges to STEM education, self-determination, and a revival of their language and culture.

At the Diné Maker Faire, there was a feeling that their days in the back seat are ending. That this time the Navajo are driving the bus.

Two Tracks to Opportunity

Navajo Tech is a leader among U.S. tribal colleges for its dual focus on technical trade skills and advanced academic degrees; it's the only one with ABET accreditation for engineering. NTU has a sophisticated high-tech Fab Lab and strong partnerships with NASA, NSF, DOE and the national nuclear labs, and aerospace companies like Boeing.

Dr. Peter Romine heads the electrical engineering department, where he emphasizes hands-on, project-based learning. Since 2015 his students have exhibited at Maker Faires in the Bay Area, New York, and Washington,

Keith Hammond, Daniel Vandever

D.C., as well as the NASA Innovation Challenge and the Nation of Makers Conference. Peter is a lead organizer of the Diné Maker Faire and a firm believer in NTU's dual-track education.

"Given the free choice," Peter asked, "what percentage of humans would choose a technical career that requires a 2-year degree, versus needing a 4-year degree? I believe the vast majority would prefer skills and jobs that help them make a living locally. I think the Navajo could become an example of that other approach to education that equally respects technical and trade skills along with academic skills. The Maker Movement democratizes learning; you don't have to be this elite-college-ready person, you could become or do anything, make money, have a fulfilling life, create businesses, employ people."

Daniel Vandever, NTU's communications director, is another key organizer of this Faire. He exudes Navajo pride and a keen sense of irony. "My grandfather was a code talker [in World War II] who helped save the nation by speaking his language," he said. "My father was beaten for speaking Navajo in boarding school in Utah." Daniel earned his degree at NTU in Navajo Culture, Language, and Leadership and wrote a kids' book based on his father's experience; now he's calling on Navajo makers to envision their place in a prosperous future. "Our goal at NTU is to provide pathways to opportunities," he said, whether academic or technical. "We want all students to engage with academics, but know that it does lead to careers."

Four Worlds of Making

Diné Maker Faire is organized according to the Navajo creation legend of the Four Worlds, Daniel explained, to show how the art of making has evolved over time. The Black World represents the oldest, traditional knowledge and crafts; Blue World showcases trades and careers in the modern wage economy; Yellow World is creativity and reclaiming identity; White World is today's newest technology. At this year's event, a fifth, Glittering World represents youth and the future of making.

Here are some highlights from the last Maker Faire of 2020 before COVID-19 shut down the world. See more at makezine.com/go/dine.

■ Black World — Tradition

Keith Hammond

DOLL MAKING: Artist and teacher Barbara Morgan showed how to sew traditional Navajo costumes, based on historical sources and old family photos. "Before the treaty, we were loud, proud, and mean," she said. "After the treaty we dressed in rags."

SILVERSMITHING: Morgan also showed fine silver work by her son Julius K. Morgan. These beautiful belt buckles were not stamped, but cast in molds sculpted from soft volcanic tufa rock, which lends itself to thicker, serpentine forms.

MAKING A NAVAJO BUN: Undergrad IT major Shalii White (foreground) showed women and girls how to make a hair tie from loops of yarn, then use it to fold up their long hair into a *tsiiyéeł*, the traditional Navajo bun hairstyle.

WEAVING: NTU students learn to build a loom, use native plants for dyes, and weave wool rugs and baskets. Freshman IT major Christa Goodluck showed a traditional basket that will be sealed with pine pitch for carrying water.

KIDS' MAKER CORNER / CREATIVE MINDS was a fun kids' activity area making crafts and gadgets, sponsored by the NTU early learning and preschool staff. Tables were filled with kids the whole day. I noticed Forky from *Toy Story 4* next to a traditional Navajo cradleboard infant carrier.

■ Blue World — Trades

WELDING: Instructor Chris Storer led a hands-on Learn to Weld activity and a welding contest. He's also a silversmith, and he was proud to show off this awesome steel sculpture of an ear of corn created by students Gabby Bryant and Leomie Foster: "The girls are doing great work!"

AUTOMOTIVE TECH: Instructor Shawn Piechowski showed off the Polaris UTV his students are modding with a DynoJet quick-shift sensor, new clutches, Portal gear reductions for higher torque, and 24" lifts for extreme high clearance over rocks and gullies. "I've worked on a lot of these for hunters, but this one is our monster sheep-herding rig," he said. "There's a real need for this on the Navajo Nation."

Yellow World — Identity

COSTUME FABRICATION: Kirk Tom wears his handmade cosplay *Star Wars* firetrooper armor emblazoned with his original insignia painted in the style of traditional Navajo weaving. "I always like to make my own designs," he says. No CNC or 3D printing here; that helmet is sculpted entirely free-hand from EVA foam.

SHOE DESIGN: Basketball rules the rez, and Tristan ("Tris") Mexicano is a baller. "Until freshman year at NTU," he says, "I never had more than one pair of shoes." A friend taught him to take care of his new basketball shoes and they started customizing them for games. Tris is now a self-taught footwear expert, able to disassemble, reglue, and reshape modern shoes. He donates refurbished pairs to kids in need, and even takes small commissions customizing shoes for athletes and friends.

White World — Today's Tech

3D PRINTING: Jonathan Chinana exhibited his Soft Robotics experiments; he's using 3D printed molds to cast silicone rubber actuators and grippers that change shape when inflated. Jonathan did a summer research program at Harvard University and came back full of skills and ideas but lacking some of their fancy equipment. He built this DIY vacuum chamber from PVC pipe and 3D printed end caps, for degassing the silicone castings.

ELECTRICAL ENGINEERING: Student makers showed off CNC pen plotters, a massive wrap-around flight simulator, and a 3D-printed Wile E. Coyote picture, rigged with an ultrasonic distance sensor and Arduino. It makes Road Runner sound effects when it detects your presence.

RESILIENT CLEAN ENERGY: Instructor Darrick Lee demonstrated a Solar Micro Grid for powering a small group of homes. Much of the vast Navajo reservation is still without electricity, and homes are commonly miles apart. "The younger generation wants electricity," he said, "but it costs $27,000 per mile for the power company to extend a line to your house." Darrick's off-grid inverter design can also be adapted to grid-tie to a 25kV utility power line.

MECHANICAL ROPING: A robot "calf" zooms out from underneath this dummy horse for roping practice. It's a joint project of an electrical engineering team led by Hansen Tapaha, and the NTU rodeo team.

METAL 3D PRINTING: NTU Fab Lab's new Optomec metal 3D printer uses high-powered lasers to weld metal powders completely, no binders or post-curing, just solid metal parts right out of the printer! Additive manufacturing (AM) technician Gregory Dodge showed me a solid stainless steel print: a traditional Navajo stamp for silversmithing.

■ Glittering World — Youth and Future

KIDS 3D PRINT CHALLENGE: Middle and high schoolers competed in the Innoventure Design Challenge to design and 3D print a product to help Navajo elders, co-sponsored by KARMA, the Ke'yah Advanced Rural Manufacturing Alliance, a Native entrepreneurship program.

▌▌ Next-Gen Navajo

What lies ahead for technology-minded young Navajo? Maker careers are already in front of them — good-paying work in digital design and fabrication, engineering, coding, and precision manufacturing. They'll build the next-gen water and power infrastructure on the Navajo Nation, and clean up the uranium mistakes of the past. And if nuclear power is part of the world's climate solution, shouldn't the Navajo be designing and operating safer mines, mills, and reactors, to avoid repeating those mistakes? All this would've sounded impossible 20 years ago. On this day at Navajo Tech, nothing really sounds impossible. ●

Makers:
The Countermeasures for Covid-19

Of this, one day, epic ballads will be sung.
—Glenn Walters, University of North Carolina at Chapel Hill

Written by Dale Dougherty

People are part of PLAN C

DALE DOUGHERTY is founder and president of Make: Community, which produces *Make:* magazine and Maker Faire.

Underfunded, undervalued, and often unappreciated, makerspaces have happened any way possible in local communities over the last 10 years. Struggling but still surviving (with our condolences to TechShop), makerspaces can be found in schools, libraries, science centers, universities, and children's museums; the United States alone holds over 1,000 makerspaces. By the end of March 2020, however, almost all of them were shut down along with their institutions and businesses. In spite of losing access to their spaces, makers sprang into action, connecting to other makers, building on relationships they had developed through makerspaces and Maker Faires.

Credit Neil Gershenfeld, who started Fab Labs (featured in the first issue of *Make:*), with the big idea that digital designs could be shared and distributed around the world and then replicated on a variety of machines in each local area. The idea was to disrupt globalization, which locates production where labor is cheapest. By democratizing the tools and processes of production, digital fabrication could maximize the creative and technical talent in any part of the world and create products that could be adapted and made locally. Unfortunately, Gershenfeld's model, which made sense in theory, never became practice. Until now. Until rapid prototyping and digital fabrication became the solution to fighting a virus that, well, went viral, hitting us hard globally and locally at the same time.

Self-organizing efforts and a widely distributed infrastructure provided for a unique civic response of makers to Covid-19, not just in the U.S. but around the world. As Covid-19 began to spread, makers launched a series of countermeasures to address shortages of medical supplies and equipment in communities and protect frontline workers and the public from greater harm:

- Working alone at home but together on the internet, makers collaborated to **design new medical gear** that can be 3D printed or laser-cut anywhere.

- They organized expert networks to **evaluate those designs**, medically and mechanically, to be sure they'd get the job done.

- They fired up their own machines, and those borrowed from universities and makerspaces, to **produce millions of pieces of PPE** and other gear.

- And they **formed delivery networks** to connect with frontline health workers and hand over the goods.

"Plan C is the backup plan for the backup plan."

A countermeasure might be considered a short-term or temporary solution, a Band-Aid, an action taken to prevent a problem from becoming worse. As used in lean manufacturing and derived from Toyota's philosophy, *countermeasure* "implies there may be additional improvements, and the action taken may not be the final one, just the best one at this time." (leansixsigmadefinition.com)To use another term, they're *hacks* — something clever that just works, for now.

I called this maker response "Plan C." If the government's response to the crisis is Plan A and industry's response is Plan B, then Plan C is the backup plan for the backup plan. The Plan C countermeasures of makers might not be the best option if Plan A and Plan B offered better ones. Countermeasures are an appropriate response, preventing a bad problem from becoming worse.

Countermeasures can show the way to systemic changes as well. Indeed, cultural change often happens bottom-up to influence and shape the mainstream. In Margaret Mead's famous quote, the efforts of individuals and small groups are the only way change happens.

As Gershenfeld imagined, makers (whether they used that name or not), regardless of their location, background, experience, or job title, are creating digital designs for PPE and sharing them

Chicago makers created Illinois PPE — a network of dozens of makers producing thousands of face shields on laser cutters and (below) 3D printers.

Makers are creating digital designs for PPE and sharing them openly and freely, often modifying them in small but valuable ways.

A PLAN FOR RECOVERY

The civic response by makers can also be seen as a call to public service. Along with Rep. Tim Ryan (D-Ohio), I called for a new Civic Response Corps, inspired by the Civilian Conservation Corps, a government program that responded to massive unemployment brought on by the Great Depression. We need coordination among the self-organizing groups and also a way to train more people in the skills they need to participate effectively as civic responders. makezine.com/go/civic-response-corps

openly and freely. Others sort and select designs, often modifying them in small but valuable ways. Groups have formed to produce them locally using tools in makeshift home labs and utilizing networks associated with makerspaces to distribute them. Makers are listening to feedback from doctors and nurses to improve the designs, iterating quickly.

The maker response to the pandemic has been fast, flexible, and just flat-out incredible. I'm beyond proud to see the maker community rise up for this moment. It demonstrates the competence of this community and the many, many people it has nurtured over the years.

Makers have done what is needed to be done in a way that hasn't been done before. This is innovation in the wild — crazy and chaotic but impactful. Think of it as open and public R&D, outside institutional boundaries and good for all those things that don't have a business model, and won't be done by industry but perhaps should be done by government someday.

For the future, we have to make things better for more and more people by investing in people. Our society must increase our capacity for local production and create a sustainable economy. We must improve the resiliency and response of our institutions through new leadership. We must use our schools to help the next generation acquire the skills and mindset of makers so that every person can be productive. There are even bigger challenges waiting for us. More than ever, the future needs more makers.

Today, though, let us celebrate what makers did in the spring of 2020 and recognize their outstanding public service in a time of crisis. Many epic songs should be sung. ◉

Your old road is rapidly agin'
Please get out of the new one if you
can't lend your hand
For the times they are a-changin'
—Bob Dylan

Dan Meyer, Jeff Solin

When Makerspaces Propped Up the World

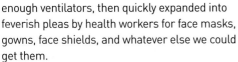

A s the Covid-19 virus spread and it became clear we were facing a global pandemic, existing infrastructure for things like surgical masks and face shields began to break down. It started with a spatter of news articles about how we didn't have enough ventilators, then quickly expanded into feverish pleas by health workers for face masks, gowns, face shields, and whatever else we could get them.

Seemingly overnight the news was filled with hundreds of stories of makerspaces using laser cutters, sewing machines, and 3D printers to deliver much-needed equipment to hospitals. Modern digital fabrication meant that designs could be shared and iterated on globally in a few minutes, and production could happen anywhere. A doctor could request a change and within the space of an email or social post, that change could be distributed around the planet.

Make: surveyed makerspaces in early April; the results showed that well over 1 million pieces of personal protective equipment had already been produced and delivered to front-line workers by makers globally. We know this response was only a tiny sample of the community that is hard at work on these efforts, and only a snapshot in time while the effort was still ramping up globally. Tallies by the OSMS network (see page 32) topped 4 million by April's end. The real numbers may never be known, but this is indisputably a moment when makerspaces have proven their community value.

Here are a few high-volume highlights from our outreach. You can still participate at makezine.com/go/ppe-survey. —*Caleb Kraft*

Sampling of Global Makerspace PPE Production, April 2020
Make: Survey April 2020, selected highlights

1. **California:** Hackerlab Sacramento & Rocklin – 1,000 face shields per day
2. **Oklahoma:** Fab Lab Tulsa — 35 ventilator splitters, 150 respirators, 2,500 face shields per week
3. **Alberta, Canada:** Fuse33 — 3,000 face shields per day
4. **Wisconsin:** Milwaukee Makerspace — 1,000 face shields per day
5. **Illinois:** DePaul University — 6,500 face shields per day
6. **Missouri:** Hammerspace K.C. — 500–1,000 face shields per day
7. **New York:** MakerSpace NYC — 5,000 pieces PPE per week
8. **North Carolina:** Charlotte Latin School Fab Lab — over 55,000 face shields
9. **Florida:** Moonlighter Miami — 1,000 face shields per day
10. **Bogotá, Colombia:** Cuadrilla Espacio and community — 5,000 face shields per week
11. **Sheffield, U.K.:** Pimoroni — 500 face shields per day
12. **Belgium:** Open Garage and community — 100,000 masks, 50,000 face shields, 10,000 gowns, 20,000 liters of alcohol
13. **Italy:** Scientific FabLab ICTP — 100 face masks per day
14. **Malta:** Invent3D — 500 face shields per day
15. **Mumbai, India:** Maker's Asylum — 1,500 face masks per day
16. **Faridabad, India:** Makershala — 1,000 face shields per day

What Are We Making?
Make: Survey April 2020, data analysis by Aiphi Wang

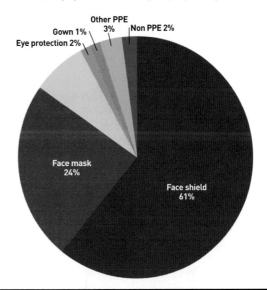

Face shield 61%
Face mask 24%
Eye protection 2%
Gown 1%
Other PPE 3%
Non PPE 2%

Facing the Infection

These makers are putting all their efforts toward Covid-19 relief

Written by the *Make:* Covid Reporting Team

Alll over the world, local endeavors are underway to provide Covid-19 relief. Many of these are viable only because of maker networks. As **Glenn Walters**, director of the BeAM (Be A Maker) Design Center at the University of North Carolina at Chapel Hill, writes:

> *Without our makerspace capacity to prototype, test, and scale production it would be unlikely that the University could have mounted a large-scale support effort for healthcare providers across the state. Similarly, the connections that have been established among makerspaces around the state have provided us with a forum for sharing problems and solutions in a way that would have been impossible ten years ago.*

Here are just a few of the efforts and initiatives by makers who are dedicated to fighting Covid-19. Find more at makezine.com/go/plan-c.

Open Source Medical Supplies Community

Robot builder **Gui Cavalcanti** (Figure **A**), known for his MegaBots project (*Make:* Volume 58), is the lead organizer of the Open Source Medical Supplies group (opensourcemedicalsupplies.org). OSMS started as a ventilator project until pivoting to PPE and broader designs.

The OSMS Covid-19 Facebook group grew to more than 70,000 in a few weeks (Figure **B**), including doctors, former Fortune 500 executives, librarians, medical transcriptionists, and makers of all kinds. The group quickly created an online "repository of vetted open source designs that fabricators around the world can make locally to provide to hospital systems in need." Moderated by 400 volunteers, the OSMS guide immediately became one of our most trusted sources of designs — reviewed by medical and engineering professionals — that makers can build, print, or sew. The OSMS network now includes 150+ local chapters around the world and has produced 4 million pieces of PPE, and counting.

A

B

Courtesy Gui Cavalcanti, Robert Read

Analysis of Open Source COVID-19 Pandemic Ventilator Projects

Look Down! We've added tabs for modules to encourage modularity!

April 15, 2020 Public Invention   https://www.pubinv.org   Home Repo: https://github.com/PubInv/covid19-vent-list

Link to definition of evaluation criteria: https://docs.google.com/document/d/e/2PACX-1vRl9yZ27KvslftcNvweHgH1A81pO8gHL62TWpY_VY-UELWdK9x-4-3hNw3DbkemClzExPsg8RfnxilP/pub

Project Name	Project Link	Openness	Buildability (1 unit)	Community Support	Functional Testing	Reliability Testing	COVID-19 Suitability	Clinician Friendly	Average	Manufactura bility (1000s)	Date Last Evaluated
Medtronic Puritan Bennett (PB) 5(	http://newsroom.medtronic.con	4	2.5	4	5	5	4	5	4.21	4	2020-04-19
Ambovent	https://1nn0v8ter.rocks/AmboV	4.5	4	4.5	4	3	4	4	4.00	3	2020-04-19
MUR (Minimal Universal Respirat	https://www.mur-project.org/	4	4	4	3.5	4	3.5	3.5	3.79	2.5	2020-04-19
Open Source Ventilator Project	https://simulation.health.ufl.edu	4	3.5	5	3.5	2.5	4	3.5	3.71	4	2020-04-19
Rice OEDK Design: ApolloBVM	http://oedk.rice.edu/apollobvm/	5	4	4	2.5	2.5	3	2.5	3.36	2.5	2020-04-19
A.R.M.E.E. Ventilator	www.armeevent.com	5	5	4	2	3	2	2.5	3.36	5	2020-04-19
COVID-19 Rapid Manufacture Ve	https://www.instructables.com/i	5	4	4	3	0	3.5	3.5	3.29	2.5	2020-04-19
OpenVentilator (PopSolutions)	https://www.popsolutions.co/en	5	3.5	4	3.5	3	2	2	3.29	3	2020-04-08
Low-Cost Open Source	https://github.com/jcl5m1/venti	5	4	4	3	1	3	3	3.29	3	2020-04-19
DIY-Beatmungsgerät [Respirator]	https://devpost.com/software/d	5	4	3	2.5	2	3	3	3.21	0	2020-04-19
PREVAIL NY	https://jmawireless.com/prevail	4.5	4	3	2.5	0	4	3.5	3.07	3.5	2020-04-19
VentilAid	https://www.ventilaid.org/	5	4	4	3	0	2.5	2.5	3.00	2.5	2020-04-08
Protofy Team OxyGEN	https://oxygen.protofy.xyz/	5	4	4	3	1	2	1	2.86	3	2020-04-08
Jeff Ebin's Prototpye	https://www.ebcore.io/?fbclid=h	5	4	3	1	0	4	3	2.86	1	2020-04-08
Open Breath Italy	https://www.openbreath.it/	3.5	2.5	3.5	3	1	3.5	2.5	2.79		2020-04-19
CoRescue	https://corescue.org	4	1	4	3.5	0	3	3.5	2.71		2020-04-19
MARK-19 Ventilator	https://www.mark-19.com	3	2	3	4	3	2	2	2.71		2020-04-19
Rice OEDK Design: (2019) Apollc	https://docs.google.com/docum	3	3	3	3	2	3	2	2.71		2020-04-19
VentCore Ventilator	https://www.ventcore.health/	5	4	3.5	2	0	2	2	2.64	3.0	2020-04-08

Public Invention's open source ventilator project rankings.

Tracking Open Ventilators

Robert Read (Figure **C**) and his nonprofit group Public Invention (pubinv.org) are compiling and systematically scoring a ranking of about 100 open source ventilator projects (makezine.com/go/open-ventilator-projects). Their independent evaluation and testing provides important feedback for designers as well as future builders.

"Building this spreadsheet," Read says, "has convinced people that this problem is 90% testing and 10% design." He and collaborators **Geoff Mulligan**, **Lauria Clarke**, **Juan E. Villacres Perez**, and **Avinash Baskaran** have created a ventilator testing strategy and a call for modular assembly designs to allow for distributed manufacturing.

"Instead of building ventilators," Mulligan says, "what people need to do is to modularize the ventilator project itself, so that your typical *Make:* magazine reader can work on a small part of the ventilator, not be responsible for the whole ventilator themselves."

The team has also designed and shared a device for testing and monitoring ventilators, called VentMon (Figure **D**); they aim to develop it as a full-featured monitor that's modular so it can be "plugged in" to emergency ventilators designed by anyone. Find it at github.com/PubInv.

MORE ONLINE COVID-19 RESOURCES: CovidBase (covidbase.com), NIH (3dprint.nih.gov/collections/covid-19-response), Relief Crafters of America (facebook.com/groups/reliefcraftersofamerica).

Facing the Infection

CHICAGO, ILLINOIS
Illinois PPE Network

Just before DePaul University closed its doors for the lockdown in March, **Jayson Margalus** (Figure **E**, left), **Eric Landahl** (Figure **F**), and colleagues "liberated" the machines in the university makerspace, the Idea Realization Lab (IRL). "We took 3D printers and dispersed them among all of my student employees, who are currently fabricating things at home," Margalus says.

Eric and wife **Sarah Rice**'s home (Figure **G**) would become the first node in a "mesh network" that Jay and Eric began to put together. "The idea is to keep extremely isolated groups sharing information, but not exchanging people or materials for safety," Landahl says. There are now 20 nodes of varying sizes; if one node gets sick, their jobs can go to another node. By April 2, Landahl's downtown node had delivered its 500th PPE. He had also prepared a shipment to go to the hospital where his 75-year-old mother works.

When Harold Washington Public Library and its Maker Lab closed, **Sasha Neri** (Figure **H**) brought two FlashForge units home with her. Other library staff brought home Makerbot Replicator 2s and Dremel 3D printers. "The library has been supportive," Neri says. "It asked for volunteers and now there are nearly 20 people sewing and 3D printing, connected to the Illinois PPE network for pickups, deliveries, and supplies." Neri also worked with **Dan Meyer**, who made the Chicago Shield (thingiverse.com/thing:4290688), a design that can be printed on smaller 3D printers, then unfolded into shape with the help of a warm oven or hot water. So far, 22 suburban libraries have connected to Illinois PPE, each one producing 20 to 80 shields a week.

Iraq vet **Ray Doeksen** was organizing veterans to deliver PPE when he remembered the 1995 heat wave that killed hundreds in Chicago's poor neighborhoods. He figured that, similarly, "The pandemic was bound to kill people unevenly and unfairly in Chicago."

Doeksen and Margalus reached out to **Jackie Moore** (Figure **I**), founder of Chicago's free youth makerspace. Moore's robotics team wanted to

Jay Margalus, DePaul University, Eric Landahl, Sasha Neri, Jackie Moore, Jeff Solin, Johnny Lee, Sabrina Paseman

print PPE but had only one 3D printer, so they had started sewing masks. She saw that Illinois PPE wasn't serving the parts of the city where most African-Americans live. But she was well connected to local grassroots organizations.

"I started mobilizing those people," Moore says. "Now we have four nodes that are serving primarily the south and west sides of Chicago." They began by assembling and packaging PPE from other nodes; they're now 3D printing their own. "We were already doing civic response, every citizen engaged, but they weren't aware of the power of making."

Flatpack Face Shields

Jeff Solin (Figure **J**) teaches computer science and runs the makerspace at Lane Technical High School in Chicago. When the school closed because of the pandemic, he got permission to use the makerspace to start making PPE.

Solin designed a face shield that could be laser-cut in 2 minutes — much faster than 3D printing — and made entirely out of a sheet of PETG plastic, without elastic straps or other components. He called it the Solin Flatpack. "You could ship a hundred of them in a small box," he says.

Solin began producing the shields on his personal Glowforge laser cutter, working with University of Chicago Hospital doctors to iterate the design (makezine.com/go/solin). When he ran low on PETG, he reached out to the Workshop 88 network, who in turn connected him with manufacturer Triangle Dies and Supplies. They're now die-cutting the Solin Flatpack face shield, for free, at a rate of 2,000 an hour.

CALIFORNIA
CPAP-PAPR Conversion

There are millions of CPAP machines in the world. With one that he uses for his own sleep apnea, **Johnny Chung Lee** (Figure **K**), a maker in the San Francisco Bay Area, thought about modifying these breathing machines to make a ventilator. (Lee contributed to the first issue of *Make:* with his DIY Steadicam project, 14dollarstabilizer.org.) He created two different devices and published his

designs on GitHub (github.com/jcl5m1/ventilator). One is a DIY BiPAP ventilator that he called "a last resort-only option" because of the risks of using it. The other is a low-cost Powered Air Purifying Respirator (PAPR), which provides filtered air inside a protective mask or suit worn by those caring for Covid-19 patients. Lee considers the PAPR device much less risky.

MacGyvering N95 Face Masks

Sabrina Paseman (Figure **L**) is one of three ex-Apple engineers who realized the N95 respirator shortage might be solved with a simple fix to

Facing the Infection

Fix the mask.

Don't wait for an N95 #fixthemask

(M)

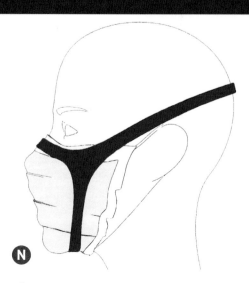

(N)

standard surgical masks involving just three rubber bands. Paseman did her own hands-on research and found that surgical masks, which are much cheaper and more abundant than N95 face masks, are made of the same melt-blown fabric, so their ability to filter tiny particles (95%) is the same.

She then discovered that the major difference between the two types of masks is the fit: the N95 mask fits tight against a person's face, while the surgical mask is a loose fit that tends to leave openings on the side. Paseman began brainstorming with fellow engineers **Simon Lancaster** and **Marguerite Siboni**. The answer, which came to Paseman, struck her as "freaking perfect." Rubber bands.

Once they'd proven the fit with a low-tech exhalation test, Paseman organized FixTheMask. com (Figure **M**) to share their DIY Surgical Mask Brace and raise funds for manufacturing a permanent version. They've brought on MIT engineer **David McCalib** to design a new V2 brace that's cut out of thin rubber sheet (Figure **N**), and they've partnered with University of Pennsylvania Perelman School of Medicine to test the fit while they await CDC/NIOSH certification.

Swim Mask Respirators

Makers with medical expertise are in high demand right now, and many of them feel called to action.

"I was a maker before I was a doctor," says **Dr. Randell Vallero**, M.D. He and his son **Connor Vallero**, a high school junior in Sacramento, created a PAPR from a full-face swim mask, tubing, computer fan, vacuum cleaner HEPA filter, and 3D printed parts, all powered by a 5V USB power bank (Figure **O**). When the fan turned out to be underpowered, they succeeded with an air pump for inflatable rafts (instructables.com/id/ COVID-Swim); Dr. Vallero has worn the PAPR all day in the OR. Next they developed a non-powered Swim Mask Respirator, using common filters

O

P

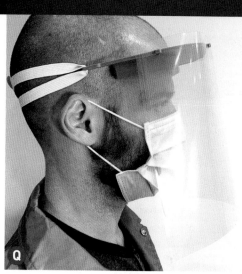

Q

like 3M's P100, HEPA vacuum cleaner bags, and hospital standard Iso-Guard in-line air filters (instructables.com/id/Covid-19-Swim).

"To patients we may look ridiculous with this protective equipment but in crisis mode, you use what you have," Dr. Vallero says. "My brother, an ER physician at a major California health system, is in the frontline in the battle with the coronavirus. Every day at work he is exposed and in crisis mode he is more than willing to use the swim mask and adapted filters, despite the fact that he feels he is wearing a 'pink *USS Enterprise*' on the top of his head."

NEW YORK
Budmen Face Shield

Two creative professionals who had started a 3D printer company, **Isaac Budmen** and **Stephanie Keefe** (Figure **P**) designed a 3D printed face shield (Figure **Q**) when their local healthcare facilities ran low on PPE. They called up the local hospital in Syracuse, New York, and asked if they could donate 50 face shields that they had made in two days. The hospital asked them if they could make 300. Two weeks later, they were making 1,000 a day in an unused soundstage in town. Volunteers showed up to help make the face shields, including a 72-year-old Vietnam vet who knew nothing about 3D printing but learned quickly to operate the machines. Their design,

upgraded with the help of **Kimberly Gibson** and **Michael Cao** at IC3D Inc., was one of the first to be approved by the NIH for clinical use (3dprint.nih. gov/discover/3dpx-013309).

NYCMakesPPE

Jake Lee (Figure **R**) is a space nerd and grad student in machine learning who thought he'd be walking in a commencement this spring, not making emergency medical supplies. But as a "superuser" in the Columbia Makerspace he was

R

Facing the Infection

already a leader with the skills to train students and staff to use the equipment, and to help them envision how to fabricate their projects.

So when NYC was hit hard by Covid-19, Lee helped organize NYCMakesPPE.com, a joint effort of local makerspaces, universities, and small manufacturers, to coordinate DIY production and delivery of PPE to health workers in need. They also launched a GoFundMe campaign to pay for supplies, overseen by Lee, **Eric Skiff** of NYC Resistor, **Guan Yang** of Hack Manhattan, and **Peter Hartmann** of Fat Cat Fab Lab. At press time they had raised over $69,000 and delivered more than 24,000 pieces of vital equipment to hospitals in four of the five boroughs.

Nurse Survives Covid, Prints PPE

At NYU Langone Health medical center in Brooklyn, **Victor Ty** (Figure **S**), a clinical research nurse, came down with mild fever, then tested positive for Covid-19 four days later, even though his symptoms had subsided. He was back to work this less than a week after that, wearing a 3D printed face shield he made at home on a 3D printer he himself had built years ago. He mentioned the Budmen design on Facebook, and fellow nurses asked him to make more. He's made 50 so far, with the help of **Matt Griffin** at Ultimaker and **Tim Leonetti** at NYU Tandon School of Engineering.

Rory Larson, Victor Ty

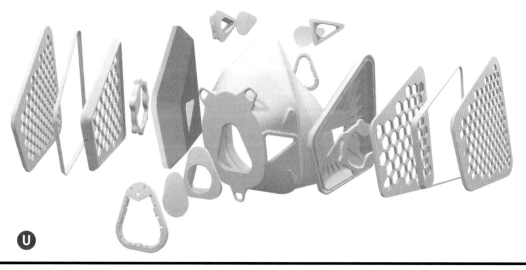

SEATTLE, WASHINGTON
The Maker Mask

A local hospital called a Seattle makerspace to ask if anyone could help out to provide them with masks and visors. A scramble ensued to figure out who could do what. **Rory Larson** (Figure , left), a maker with a mechanical engineering background, jumped in and spent all night iterating on a 3D printed design for a respirator mask using replaceable HEPA filters (Figure). "In a night I produced my first Maker Mask," Larson says.

After working until 5 or 6 am, Larson called his dad to say "Hey, look what I made." His father, **Garr Larson** (Figure T, right), a tech entrepreneur, thought it was interesting enough to reach out to his network. "Rory was insistent that the design is open source and free to everybody," Garr says.

At Seattle Children's Hospital, they reached out to Dr. Xuan Qin in the Microbiology Lab, which had run out of masks. She tested the mask, showed it to her staff, and gave them a thumbs-up. "They honestly just loved it," Garr says. "It's reusable and takes much less filter material, which is really the hard thing to find."

The mask isn't intended for doctors working directly with patients. "Because it is all plastic, it is thicker and it muffles sound," Garr says. But doctors aren't the only ones who need masks. "It is for non-patient facing jobs, for clinicians, not for ER doctors, and beyond that also policemen and firemen, grocery workers, Amazon delivery workers."

DETROIT, MICHIGAN
UAW and Ford Make Face Shields

At Ford Motor Company, a design and prototyping team inhabiting the former TechShop Detroit in Dearborn found an online design for a face shield they liked from UW–Madison and tested some prototypes. The former GM of TechShop, **Will Brick**, is on the team, as well as TechShop founder **Jim Newton**, working from Palo Alto. **Jeff Sturges**, who started several makerspaces in Detroit, one of them in a church, is the team lead. Various departments at Ford came together

and soon UAW workers in Troy were working in teams to assemble face shields. Over 500,000 face shields have been made so far.

TENNESSEE
Laser Cutting Face Shields

Bill Hemphill (Figure), an associate engineering professor at East Tennessee State University (ETSU) found a 3D design for a face shield on opensourceventilator.ie and then, like Jeff Solin in Chicago, modified it to produce more quickly on a laser cutter. Hemphill made some demonstration units and showed them to the university's director of emergency management, Andrew Worley, who told him that ETSU could start making them for the State of Tennessee.

"We banged out 500 shields in 10 hours," Hemphill says. He got some assistance from another faculty member in a different program who volunteered to come in because he "was going absolutely stark crazy in wanting to do something." Hemphill taught him how to operate the controls on the laser cutter in five minutes.

The university president, Dr. Brian Noland, suggested that they set up production in the football locker room, noting that the team wouldn't be practicing any time soon. "Given the need, with volunteer staffing for 24/7 production, this could scale up easily to 2,800 shields a week," he says.

Facing the Infection

CZECHIA
Print-Farming Face Shields

Josef Prusa (Figure **W**), whose machines have won *Make:*'s 3D printer rankings three years in a row, decided that replicating untested designs for ventilators wasn't the best place to start for 3D printing hobbyists who wanted to help with the Covid-19 crisis. His company rapidly developed a printable design for a protective face shield for medical professionals (Figure **X**) that was reviewed by the Czech Health Ministry.

Josef put his company's 3D printing farm — the world's largest — to work printing shields and shared the design openly (prusa3d.com/covid19). It became one of the most-produced designs in the world, adapted by other makers for laser cutting and injection molding as well (see page 42). And because PETG plastic works better than PLA for PPE (it's tougher and more heat-resistant) Josef is selling discounted orange PETG filament to help makers produce shields affordably.

ITALY
DIY Ventilator Valves

Italian engineers **Cristian Fracassi** and **Alessandro Romaioli** (Figure **Y**) put their skills to use when their local hospital ran out of ventilator valves, which must be replaced for each new patient. They designed and 3D printed a working rendition that became international news as the world began to recognize that governments and corporations weren't equipped to satisfy the spiking demand for medical equipment. The duo also designed and shared a valve for snorkel mask respirators at isinnova.it/easy-covid19-eng.

SPAIN
Coronavirusmakers.org

César García Sáez (Figure **Z**), co-founder of Makerspace Madrid and producer of Maker Faire Madrid and the La Hora Maker weekly newscast, is a core member of coronavirusmakers.org. This group has been instrumental in coordinating all stages of production and distribution of PPE to medical and frontline workers

Courtesy Prusa Research, Issinova, Coronavirusmakers.org, Akiba

around Spain. César has worked directly with municipal and government leaders to verify and streamline designs and distribution.

JAPAN
UV-C Sterilization

Two hours outside Tokyo is Hackerfarm, where you'll find Christopher Wang (Figure ), better known as **Akiba**. He was a major contributor to the Safecast project, which formed when the Japanese government failed to provide information about local radiation levels after the Fukushima disaster. The Safecast team designed and deployed handheld Geiger counters to measure radiation levels in real-time.

Now, Akiba is building low-cost, open source decontamination units that allow for safe N95 mask reuse. Called Hyjeia, after the Greek goddess of cleanliness, the project began with a UV-C light box and a dosimeter circuit that measures the UV so users can verify that it works (hackerfarm.jp/projects/hyjeia). Next, he scaled up to a 80-watt, wall-sized decontaminator similar to commercial units. "We're trying to use open source hardware and software to bring the cost of the equipment down from around $50,000 to around $100 to $200," he says. He sees them being used outside of hospitals where supplies of masks are limited. Once again, Akiba finds himself measuring radiation levels, although at much safer levels. ⊘

Aa

DIY PPE Rockstars

Face Shields

These are primarily being 3D printed, laser cut, die cut or stamped, and injection molded.

Most 3D printable face shields consist of a simple frame; the original styles were open on top, but some later incorporated a top barrier — a requirement for some medical facilities, although slower to print. Approaches range from simple headbands with hooked ends in the back for elastic or string, to fully printed, adjustable wraparound units, to types that sit on the ears and nose like glasses. A sheet of clear plastic is clipped onto nubs or grooves on the outside of the frame, wrapping around the wearer's face. 3D printing offers the most customizability, but at a slower speed. Prusa (page 40), Budmen (page 37), and 3DVerkstan (3dverkstan.se/protective-visor) are popular models with NIH approval.

Laser cutting is much faster. Largely using just clear PET sheets, designs range from simple all-in-one units like Jeff Solin's (page 35) and Helpful Engineering's (helpfulengineering.org/projects/origami-face-shield), to multi-piece assemblies like those from Bill Hemphill (page 39) and Protohaven (protohaven.org/proto-shield, shown above, left). These can produced in minutes, or even seconds on the simpler styles.

Injection molding, stamping, and die cutting have the fastest output by far once the tooling is created, but they're expensive to modify after a design is set — part of what made the first wave of 3D printed designs so valuable is that they allowed for quick iterations to determine viable designs for industrial approaches. Makerspaces, some working with local small manufacturers, are now creating tens of thousands of shields per day with these techniques.

With all face shield approaches, rapid design work has interplayed with innovative solutions to problems like unexpected material shortages (filaments, elastics, etc.), printer size restrictions (resulting in unfolding mask designs, among others), and the need for speed (prompting slimline and stackable visors).

Mat Thorne

Face Masks

In addition to the splash-resistant surgical masks requested by health workers, plain fabric masks are currently CDC-recommended for everyone, to contain coughs and sneezes. The OSMS group has approved several designs for DIY production, including the MakerMask Surge (shown here), made of nonwoven fabric from grocery bags (makermask.org/masks/surge). Local groups, like those affiliated with Relief Crafters of America (facebook.com/groups/reliefcraftersofamerica), are connecting with local medical professionals to get the right projects done and into the right hands. Some hospitals put out kits that include all materials you need to produce masks; after assembly, a designated collector picks them up.

As has become common with Covid-19, many general materials for these masks are now in short supply. Makers are producing novel solutions in response. Simple 3D-printable bias tape jigs (thingiverse.com/thing:4232886) have become a useful tool for helping fold over fabric strips used as ties. The CDC has even circulated a diagram of how to fold a bandana with a coffee filter inside as a makeshift option (makezine.com/go/bandana-facemask).

3D printing has also been useful for making "ear savers" (thingiverse.com/thing:4249113) that connect a surgical mask behind the head, rather than around each ear (which can result in pain and injury after hours of use).

Laser cutters are being used with freely distributed templates (makezine.com/go/laser-face-mask) to cut large batches of mask profiles, allowing for quicker assembly.

Groups and individuals are also examining the properties of various materials that can be used for face masks and respirator filters. Camron Blackburn is documenting her filtration research at MIT, even electron-scanning different options to determine which work best (camblackburn.pages.cba.mit.edu/filter_media).

DIY PPE Rockstars

Respirators

Like face shields and masks, N95 respirator masks too have become scarce, and the 3D printing community similarly has responded with designs for printed versions.

Experts have indicated maybe this isn't such an easy fix. Printed masks present real challenges:

- Plastic from your standard filament-based printer is porous and difficult to sterilize. It's generally not recommended for medical uses because it harbors bacteria.
- The rigid nature of 3D prints makes it difficult to ensure a proper seal to your face. And if a respirator mask isn't sealed to your face, what good is it anyway?

Some in the community have warned that a 3D printed mask is a liability. "If you absolutely insist on printing a mask now, treat it like it is a basic surgical mask and not as a true respirator with all the protections they provide," advises Josef Prusa. "A false sense of security can be very dangerous." Others respond that even a non-perfect mask is likely better than nothing.

Some medical facilities have now begun accepting certain 3D printed masks. These have not gone through the NIH approval process, but local hospitals have assessed them and decided to give them a try.

The Montana Mask (makethemasks.com) is being printed in homes and businesses and used at hospitals in Montana, Texas, California, and other states. This simple design (shown above) allows for a replaceable filter material (a key aspect), and uses a gasket made of rubber window weatherstripping to provide a seal to your face. The MakerMask out of Seattle is another 3DP mask (see page 39) but it's more complex.

Since his early warning, Prusa has published scientific testing to show which methods work (and which don't) for sterilizing prints made of PETG. Technically you can sterilize a 3D printed mask. The Covid-19 Healthcare Coalition offers their own insights into PPE production and sterilization at c19hcc.org/insights.

Throughout April, we've seen new respirator designs appearing every day, including a simple mold for casting respirators out of silicone at home and an NIH-approved 3DP mask, developed by 3D Systems and the Veterans Health Administration, using a fancy SLS nylon powder printer for medical use. And, in a sign of where all 3DP masks may be heading, the Montana Mask has been adapted for injection molding.

—*Caleb Kraft and Mike Senese*

makethemasks.com

MAKE, FIGHT, WIN

You too can produce PPE and other Covid-busting gear! Make all these projects, and many more, right now at makezine.com/go/the-big-list.
—Keith Hammond

Hand sanitizer, official WHO recipe

HAND SANITIZER
Keep those paws clean, cats and kittens.

WHO-recommended Hand Rub — It's just glycerol, hydrogen peroxide, and lots of ethanol or isopropyl alcohol. For a big batch, see who.int/gpsc/5may/Guide_to_Local_Production.pdf. For a home-sized batch, see youtu.be/I-e9AEaz0FQ.

PROTECTIVE SUITS AND GOWNS
Full-body protection for workers at the front lines.

PS-1 Open Source Protective Suit — Michelle Dulce in Manila reverse-engineered an isolation suit and created a sewing guide; Alex Crease from Boston created DXF files for machine cutting. OSMS says use Tyvek 1433R fabric and follow DuPont's method of heat-sealing the seams.

PS-1 SUIT			STYLE IMAGE REFERENCE
STYLE NUMBER:	PS-1 SUIT	FABRICATION	SPUN-BOND FABRIC
DESCRIPTION:	REPLICATED PROTECTIVE SUIT	INITIATED BY:	MICH DULCE for MANILA PROTECTIVE GEAR SEWING CLUB
DATE:	MARCH 25, 2020	CREATED BY:	KENDI MARISTELA

FRONT　　　　BACK　　　　SIDE
PS-1 Open Source Protective Suit

STERILIZATION
Sanitize and reuse protective gear.

YouVee: A DIY UV-C Irradiation Cabinet — With $50 and one trip to the home improvement store, build this ultraviolet light cabinet to zap the virus and sterilize N95 masks and other gear. Easy project uses a germicidal UV-C bulb, an ordinary work lamp, and aluminum foil tape. Developed by Deeplocal in Pittsburgh, Pennsylvania; build it at makezine.com/projects/build-a-diy-uv-c.

YouVee: DIY UV-C Irradiation Cabinet

EMERGENCY VENTILATORS
Inflate/deflate lungs of patients unable to breathe.

AmboVent Emergency Ventilation System — Designed by Israel's national emergency service, military, and leading hospitals, the AmboVent was open sourced on Github on April 1 and is rated one of the best open source options by Robert Read (see page 33). It compresses a manual bag valve mask (BVM) automatically using an Arduino Nano, SparkFun pressure sensor, and snowblower motor, and can be built in a typical makerspace for about $500. ⏼

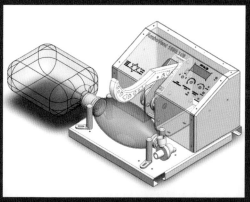
AmboVent Emergency Ventilation System

Homeschool How-To

A quick primer for all the families now educating kids at home

Written by Bob, Christine, and Jonathan Pappas

In August of 2016, our three kids Jonathan, David, and Sophia were enrolled in three different schools with three different schedules. Our evenings and weekends were spinning with everyone moving in different directions. By May, we knew we were ready for a change. We had been homeschooling one child already, but not satisfied with our approach. After attending an experiential-learning workshop, all kids excitedly jumped on board with homeschooling!

At first, my husband Bob and I began filling up their schedules with exciting classes, activities, and sports, giving them little time to breathe and work on their passions. After a couple of months, we realized this was a mistake. With the help of our mentor, we created a homemade plan for our children's education. After identifying interests, talents, and skills, we went to work locating mentors, resources, and classes to help them gain needed skills. We also protected the time they needed to be makers.

Our homeschool has not only been filled with a variety of educational learning experiences, but also and equally important — our days have the much needed time to be thinkers, dreamers, idea-discoverers, creators, makers, and world changers!

For everyone that's been suddenly thrust into this position: You have been given extra time with your children. Make the most of it! This is a chance to look for the blessings in this extra time and be explorers and creators. To think outside the box and have "aha" moments. To be the producers, be the builders, be the makers.

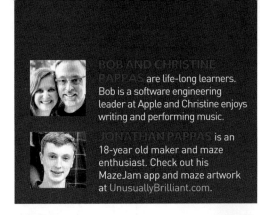

BOB AND CHRISTINE PAPPAS are life-long learners. Bob is a software engineering leader at Apple and Christine enjoys writing and performing music.

JONATHAN PAPPAS is an 18-year old maker and maze enthusiast. Check out his MazeJam app and maze artwork at UnusuallyBrilliant.com.

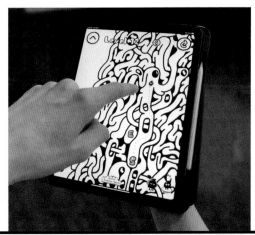

Pappas family

Sophia sews, crochets, and writes novels.

Jonathan and Sophia working on vocabulary.

David creates clever comics and games.

Pappas family

Jonathan's maze artwork masterpieces.

These steps will help you build a good plan with your family:

1. Have your children create a list of dreams and passions
- What am I passionate about?
- What makes me happy?
- What are unique skills and talents I have that I love to do?

2. Dig deep and discuss each passion
- Why is this passion worth pursuing?
- How can I develop this passion?
- What do I need to learn to see this passion grow?

3. Support transformational learning
- Discover mentors who stretch your children
- Use resources that spark "AHA" moments
- Provide plenty of time for skill building and exploration of new skills

4. Encourage an entrepreneurial spirit
- Consider how your talents, your passions can be shared with others and positively transform others
- Develop a plan to implement entrepreneurship

5. Review and reflect on transformational learning with these questions
- What happened? (concepts, activities)
- What did you learn? (the "AHA" moment)
- How did you feel about what you learned?
- What are your next steps?

Helpful resources:
- The Clifton Strengths Assessment — helped us gain a better understanding of each child.
- *Homeschooling Through High School* seminar
- Google Scholar, for research on various topics
- The Planner Pad — undated spiral-bound organizers are great for project management.
- Mentorships with teachers who have expertise in a particular field of study.
- David recommends Do Ink Animation (favorite animation app) and iFontMaker (favorite font creating app.)
- Sophia recommends the books *Stitch by Stitch* and *Made by Me* by Jane Bull, and *Super Easy Amigurumi* by Mitsuki Hoshi.

I chose to officially start my homeschooling career in 10th grade. My parents have been an amazing support, and I am grateful every day that they blessed me with this opportunity. Looking back at these last three years, it has been the best decision we have ever made.

Thanks to our mentor Mrs. E, doctorate in education, I've gathered the following helpful notes for other students trying to get started with homeschooling and moving into adulthood:

ENCOURAGEMENT FOR BEGINNERS

What are you interested in? What are you good at? What do you love to do? What do people recognize you for? You may think that since you are young, you are less important. This is furthest from the truth. Now is your time to shine and make a difference: Not after college or retirement. This is your most active time in your life. You have the most drive to do things from ages 16–24.

VISION CASTING

Where do you want to be in 5 years? 10 years? 20 years? 70 years? What do you want to be known for? When you are old, and think about what you did in your life, what will you be proud of? Plan a vision for your life, be organized and well thought out.

POST-LECTURE COMMENTARIES

Sometimes, you might be in a lecture environment. How do I make it stick and not lose it? Take notes in class, and afterwards (within 24 hours) write a commentary. Take no more than 5 minutes to answer these questions:

- What happened?
- What did I learn?
- How do I think/feel about it?
- What of it? (Why was it useful?)
- Commentaries turn lectures into experiential learning.

A few additional personal insights:
Getting stuff done

Let's say you have something you need to get done that isn't particularly interesting to you. Homework, an essay, a house chore, anything. Want to know a secret method to be your most

efficient? Try the Pomodoro method ("tomato" in Italian). Set a random timer for yourself between 16–20 minutes (the time must be random). Then, work as hard as you can until the timer goes off. Breath: 5 beats in, 6 beats hold, 7 beats out. Take a break for five minutes. Go outside. Relax. Start again, if you like. That's it! Do as many as you want in a day.

Get enough sleep

Your mind needs sleep. Sleeping heals your body. Sleeping before midnight also turns short term memory into long term memory. If you don't get 8–10 hours of sleep, the things you learned yesterday may be forgotten.

300 good decisions

Your earliest hours in the day are most efficient for your brain. As soon as you wake up, your brain has about 300 quality decisions before you are maxed out and put into autopilot (~noon)

Your early hours are for *learning tough things*. This is the best time of day to grasp difficult concepts. Your afternoon hours are for making. This is when your creative brain gets going. Don't make tough decisions at night. Tough decision + over-tired = sleep on it.

Relax

Make a day for yourself where you aren't working or doing school. Relax, enjoy life, go outside, exercise, play a game, read a good book. Taking a break is a good thing and will make you even more efficient.

Some personal favorites:
- Programming teacher: Bob the Developer
  bobthedeveloper.io
- Math teacher: funmathclub.com
- History, Literature, and Rhetoric teacher:
  balm.pathwright.com

Recommended books to kickstart your homeschooling adventure:
- *The Romance of Education* by Robert Welch
- *How to Read a Book* by Mortimer Adler
- The Great Books, compiled by Mortimer Adler
- *Word Clues: The Vocabulary Builder* by Amsel Greene
- *Oxford English Dictionary*

MIKE SENESE is the executive editor of *Make:*. He's hosted TV series on Discovery and Science Channel.

Community Broadcast Services

Watch and create your own virtual classes and webinars with these tech tools

Written and photographed by Mike Senese

We're all homeschooling now, but that doesn't mean we're stuck to a one-room prairie schoolhouse — we've got powerful technology tools at our disposal that let us partake in everything from live-streamed local classes to global webinars. These even make it easy to set up our own broadcasts so we can teach others, or just host a virtual show-and-tell or happy hour with friends and family. And we can even pull in multiple cameras using our existing smartphones.

We've used a number of these apps at *Make:*. Here are our notes on what to check out.

Video Conferencing / Webinars

GOOGLE HANGOUTS (hangouts.google.com) Launched in 2013; became a common tool for group video conferencing over the next few years. Still very frequently used, but appears to be dipping in popularity.

- **Features:** You probably have a Google account already, so likely don't have to sign up for it.

- **Hangups:** Works best for smaller groups. No webinar option or call recording options. No "see everyone" gridded view.

ZOOM (zoom.us) — Because of Covid-19, large numbers of schools have moved their classes to this platform, pushing it to become the 6th most popular site in the U.S. It's a pretty easy system for doing video conference calls, and paid versions allow for 100–10,000 attendee webinars. Despite its popularity (or perhaps because of it), many people regularly point out ongoing security and privacy issues.

- **Features:** Video recording capability makes it easy to share conversations after they've concluded. Offers gridded "gallery" view. Decent host moderation tools. The virtual backgrounds are silly fun, but can be used effectively.

- **Hangups:** Requires sign-in. Free accounts capped at 40-minute meet time, with a 100 user limit (1,000 for business accounts). Webinar accounts require a separate monthly subscription (costing $40–$6,490/month,

Make: edit team in our weekly Google Hangout.

Zoom virtual backgrounds can be entertaining.

CodeJoy uses Zoom to set up their Maker Camp e-learning broadcasts.

Adove Stock - Boyko.Pictures

EpocCam lets you use a smartphone as a second webcam.

YouTube Live is an easy platform for webinar broadcasts.

Jitsi on the monitor and EpocCam'd phone on selfie-stick overhead boom.

depending on how many attendees you'll allow). Ongoing struggles with privacy concerns and violations. Also, be careful of unwanted guests "Zoom bombing" you.

- **Tips:** When in a conference or class, set yourself to mute, then hold down the spacebar whenever you want to speak. If you have a second camera, you can quickly switch between them with a keyboard shortcut.

JITSI MEET (jitsi.org) — An open-source Zoom alternative that's quickly growing in popularity. We've had some success with it but have also found periodic failures, with guests unable to sign in. Still, a good one to play with. Check their github for even more options: github.com/jitsi

- **Features:** No registration needed; guests can simply join the meeting with a link. No time limit on calls. Open aspect means you could theoretically configure this for your specific needs.

- **Hangups:** Works best for smaller groups. No proper webinar option — users are routed through YouTube for additional needs. Call recording quality is moderate. Seems to have a few extra hiccups compared to more established systems.

- **Tips:** Like Zoom's virtual backgrounds, the "blur background" option will save you from having to tidy up before a call. And be careful when naming your conference room; this make things easy but the names save across the platform, so two different groups might end up in the same place simultaneously.

Live Streaming

YOUTUBE LIVE, FACEBOOK LIVE, PERISCOPE, INSTAGRAM LIVE — Each lets you broadcast to your followers directly from your phone. Makes mobile streams so easy! They work great, although the flexibility is limited for adding any extra touches.

RESTREAM (restream.io) — Transmit on multiple social platforms at the same time; consolidate comments and views. Free. Browser-based.

Additional Tools

OBS — "Open Broadcaster Software" is another open package that's big and powerful and immensely configurable. Want to create pro-grade online events with graphics, overlays, picture-in-picture, configurable switching, and more? This will do all that, and then some.

- **Features:** Free. Lets you bring in multiple video sources. Audio mixing. Outputs to most streaming services.

- **Hangups:** Connecting to Zoom for broadcasting requires convoluted workarounds. Some extra features cost money — using your iPhone as another camera source, for example, is $16.

- **Tips:** Make sure you need this level of video control. If you're just trying to do some distance learning, you can probably get along without it.

OBS: picture-in-picture, overlays, transitions, and more.

EPOCCAM — Turns your iPhone or Android into a camera source for streaming services, which is good because suddenly all the USB cameras are sold out. We're using this as an overhead camera for streaming builds with an old iPhone 5s.

- **Features:** Free. Lets you stream wirelessly through your computer.

- **Hangups:** Zoom's latest security updates have made this non-functional. (Thankfully we still have the old version installed.) Perhaps they'll have fixed it by the time you're reading this.

- **Tips:** Important — restart your computer after you install the EpocCam driver to get it working. And you need to have a streamable app open on your computer to get the camera to respond. ⊘

Adove Stock - Boyko.Pictures

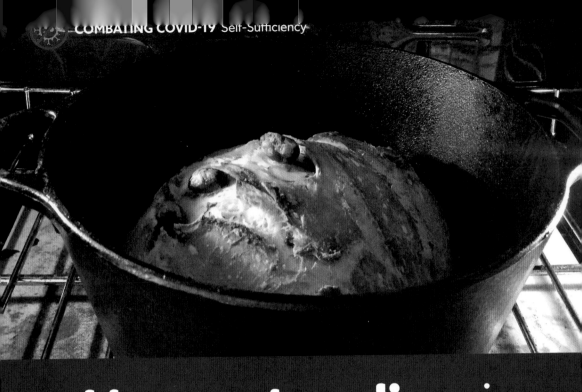

Homesteading is the New Normal

Written by Mike Senese

Brush up on your self-sufficiency skills to overcome staple shortages

Mike Senese, Adobe Stock - Stokaji

Perhaps that prairie lifestyle appeals to you more than going all-in on screentime during our lockdown. We get it! There's a huge range of skills and builds you can work on to become self sufficient in these times. Here are a few that we've focused on over the past few months. You can find steps and video examples of these and more at makezine.com/go/home-sufficiency, and if you're looking for even more resources, check out homestead.org.

Kitchen
- **Make sourdough yeast:** Tastier than the dry stuff, and not sold out in stores — you make it from the air around you.
- **Build a proofing box:** Keeps the yeast in your dough active at just the right temperature so you can pop a perfectly risen loaf into the oven for best baked-bread results.
- **Bake some no-knead bread:** Mark Bittman's recipe inspired countless variations for a good reason — it's dead simple and works great.
- **Make vinegar:** Some apples, sugar, water, and a bit of time and you're ready to go.
- **Pickle everything:** Add a tangy flavor and longevity. Try fermentation too.
- **Malt grains** for beer making and more.

Garden
- **Make newspaper planting pots:** Once your seedlings sprout these go right into the ground.
- **Plant some fast-growing veggie seeds:** Leafy greens and root vegetables are good options.
- **Build a simple, low-cost greenhouse:** You can make a tabletop version for a few bucks.
- **Assemble self-wicking garden tubs:** Makes watering less of a chore.
- **Do a hydroponic garden:** These grow large quantities of vegetables quickly.
- **Get started on a proper victory garden:** A full setup will give you all the produce you need.

Hygiene
- **Determine your no-toilet-paper options:** Bidets, used dryer sheets, cloth, recycled newspaper. Yeah... not a fun one.
- **Make your own soap:** Washing your hands is considered the best way to combat virus transmission.
- **Build a greywater system:** Water shortages are a separate issue from Covid-19, but you might as well jump on this while you're sheltering in place.

Activities and Education
- **Find and read** the Foxfire school books and Scouting handbooks.
- **Obviously, keep making stuff.**

Free Land!
The U.S. federal government no longer offers land for building a home and farm as it did with programs like the 1862 Homesteading Act. However, various rural counties and municipalities looking for new residents to help provide an economic boost do have free plots for those willing to construct and occupy a home. Many of these, like the program in Elwood, Nebraska (elwoodnebraska.com/freelots.html), are in the central portion of the U.S.; if you're willing to relocate, this is one way to score yourself a place to get going on your own personal homestead. ◐

Pump Up the Jams

Zap coronavirus with Scrubber — a Spotify-playing, soap-dispensing hand-wash timer

Written by Adnan Aga and Taylor Tabb

 ADNAN AGA is a software engineer developing novel interactive experiences at Deeplocal. Born in Dubai and now living in Pittsburgh, Adnan can be found bouldering outside or performing improv at a local theatre.

 TAYLOR TABB is an integration engineer at Deeplocal, working at the interface of hardware, software, and design. He loves weird robots and hybrid fruits, and is never far from an Arduino. Originally from Los Angeles, he graduated from Carnegie Mellon with a bachelor's and master's in mechanical engineering.

Social distancing at home but wishing you were getting down at the club? Feeling like that 20 seconds of lathering is the longest 20 seconds of your life? Looking for a DIY project to do in quarantine? Scrubber's got you covered.

Scrubber is your hand-washing soundtrack — 20 seconds of music selected from your most-played Spotify tracks of the week. When you press down on the soap pump, you'll hear your favorite jams coming right out of the dispenser as a way to time hygienic hand-washing. All you need is a Spotify account, a suitable soap dispenser, and the electronic parts listed here.

We built Scrubber for our creative agency, Deeplocal, in Pittsburgh, Pennsylvania during lockdown in March 2020. It's the first project of our "Cabin Fever" series — maker projects to inject a little bit of joy into our lives.

How It Works

Scrubber uses Spotify's API wrapped in a custom Node script. The script authenticates with Spotify, downloads and processes tracks, then plays audio through the speaker bonnet (Figure **A**) using ALSA sound software in the Linux kernel.

Don't worry, we've written the steps and the code so anyone can follow along. Before you get started, you should have some basic knowledge of command-line Linux and soldering. The Github repo (github.com/Deeplocal/scrubber) contains all the code and documentation you'll need.

Build Your Scrubber

Let's get down to it. Start by flashing a recent version of Raspbian and connecting your Pi to your wireless network. If you don't know how, there's a great guide at raspberrypi.org/documentation/configuration/wireless/headless.md. Make sure you can SSH into your Pi and can find your Pi's local IP address.

TIME REQUIRED:
4 Hours

DIFFICULTY:
Easy/Intermediate

COST:
About $30

MATERIALS

Most of these parts can be swapped for what you have access to — the speaker bonnet can be replaced with a similar audio pHat or AIY hat, and the copper tape can even be replaced with aluminum foil and super glue.

» **Raspberry Pi Zero W single-board computer**
» **Adafruit Speaker Bonnet** #3346, adafruit.com
» **Speakers (1 or 2)** We used Adafruit #1669 mini speakers to fit in our soap dispenser.
» **Thin wire, 1–2ft total length** 22 AWG or similar gauge
» **Copper tape** a few inches
» **Battery pack (optional)** We used a 5V 500mAh LiPo, Adafruit #258, with charger #1944.
» **Soap dispenser** Convert an existing dispenser, or use this nice wooden one we found at Target, #50364983.

TOOLS

» **Tiny screwdrivers**
» **X-Acto knife or scissors**
» **Soldering iron**
» **Wire cutters**
» **Wire strippers**
» **Tools to modify your soap dispenser** scissors, drill, etc.
» **Computer with internet access**

Taiyor Tabb, vecteezy.com

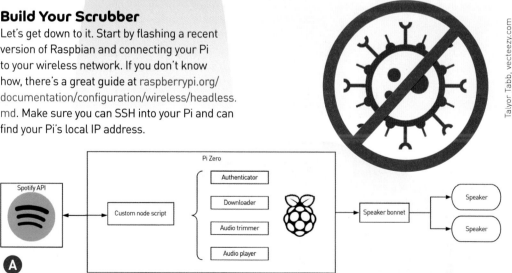

A

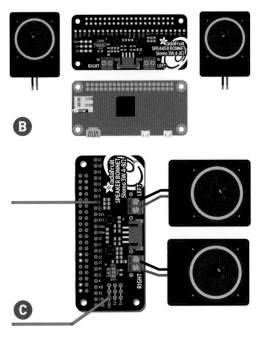

HARDWARE

1. Power down your Pi.

2. Press the Adafruit Speaker Bonnet onto the Pi's headers (Figure **B**).

3. Connect your speakers to the screw terminals on the Speaker Bonnet.

4. Locate Pin 5 and Ground on the Speaker Bonnet, and solder about 6" of your thin wire to each (Figure **C**).

5. Optionally, attach a battery pack — we did this but you definitely don't need to.

SOFTWARE

1. On your computer (not the Pi), head to the Spotify dashboard, developer.spotify.com/dashboard.

2. Click on Create an App, call it *Scrubber*, and set the description to *Scrubber*. On the checklist, select Speaker, and then click Next. Assuming you're just using this at home, click Non-commercial, read the terms, check the boxes, and hit Submit. You'll be sent to your Scrubber app's dashboard page.

3. Note down your Client ID and your Client Secret (you'll need those in a bit!).

4. Click on Edit Settings, then add http://<*Your Raspberry Pi's IP*>:5000 to the redirect URIs. So if your Pi IP was 192.168.1.232, you'd enter http://192.168.1.232:5000 and hit Add, then Save.

5. SSH into your Pi, and create a directory to house Scrubber in: **mkdir Projects**

6. Navigate into the directory: **cd Projects**

7. Set up the Speaker Bonnet with Adafruit's installer:

```
curl -sS https://raw.githubusercontent.
com/adafruit/Raspberry-Pi-Installer-
Scripts/master/i2samp.sh | bash
sudo reboot now
```

Then do it again — yup, you do actually have to run the script and reboot twice!

8. Test out the Speaker Bonnet:

```
speaker-test -c2 --test=wav -w /usr/
share/sounds/alsa/Front_Center.wav
```

then hit Ctrl+C to exit. If you don't hear audio, check Adafruit's documentation for troubleshooting at learn.adafruit.com/adafruit-speaker-bonnet-for-raspberry-pi/raspberry-pi-usage.

9. Ensure you're in your *Projects* folder, and clone this repository:

```
git clone https://github.com/Deeplocal/
scrubber.git
```
10. Install Node and NPM (we're using version 10.15.2, but any recent version should be fine).
```
curl -o- https://raw.githubusercontent.
com/nvm-sh/nvm/v0.34.0/install.sh |
bash
nvm install node
```
11. Enter the *scrubber* directory: **cd scrubber**
12. Install the ffmpeg packages.
```
npm i ffmpeg
npm i fluent-ffmpeg
```
13. Install all the relevant packages: **npm i** This may take a while. (Consider washing your hands while you pass the time.)
14. Start the script by typing **node index**, then wait for the following instructions to appear in your command line to link your Spotify account:
a. Paste your Client ID from earlier, and hit Enter.
b. Paste your Client Secret from earlier, and hit Enter.
c. In your browser, go to http://*<Your Raspberry Pi's ip>*:5000 and hit Log In With Spotify.
d. Copy the code from your browser and enter it into your SSH session when requested, then wait (several minutes) for the download to complete. You'll hear a 20-second clip play when it's all been processed.
15. Test the system by touching the Pin 5 and Ground wires together. You should hear 20 seconds of audio!

RUNNING ON BOOT
You can optionally set the script to run on boot. This way you don't have to SSH into Scrubber every time you power up.
1. Edit the file */etc/rc.local*:
```
sudo nano /etc/rc.local
```
Just above the **exit 0** at the end of the file, add this line: **cd /home/pi/Projects/scrubber && node index.js** . Then save and close the file.
2. Reboot your Pi with **sudo reboot now** . Then test your configuration by touching the Pin 5 and Ground wires. Enjoy those 20 seconds of music!

PUTTING IT ALL TOGETHER
There are a million different soap pumps out there. Here's a simple way to trigger your music that can be adapted to almost any soap dispenser.

You'll use copper tape to make the pump act just like a momentary pushbutton switch.

1. Cut two small strips of copper tape and place them on the pump just such that they make contact only when the bottle is pressed.

No copper tape? No problem. You can use glue and aluminum foil to achieve a similar effect.

2. Gently solder those Ground and Pin 5 wires to the copper tape (Figure **D**). Now, when the two copper strips make contact, the Pi will perceive a button press.

Because our workshop was closed, we couldn't make a custom dispenser, so we went to Target and picked one we could easily cut into, with space for the electronics and speakers inside. We Dremeled off the bottom panel to reveal an interior plastic soap bottle. We gutted it and affixed a smaller bag of soap inside. Below that we press-fit a sheet of plastic to support the bag. This made a compartment at the bottom just large enough to isolate the electronics.

SCRUB FOR VICTORY!
And that's it, you've got your very own Scrubber. Smash that pump for 20 seconds of hand-washing dance party energy.

Making During a Pandemic
Scrubber was concepted, designed, built, and launched the week of March 16, 2020. This presented new challenges because Deeplocal staff were working from home. We've attempted to use parts you may already have on hand. If you do find yourself ordering parts, please be mindful of the increased demands many suppliers and delivery services currently have on them. ●

Talyor Tabb

Low-Fidelity Prototyping Cart

MATTHEW WETTERGREEN is an associate teaching professor at the Oshman Engineering Design Kitchen at Rice University where he teaches engineering design and prototyping.

Addis Ababa Merkato

Sheltering in place? Create a collection of common materials to help you keep on making

Written and photographed by Matthew Wettergreen

That day my task was simple: to collect wood and plastic. I needed these materials to teach engineering faculty how to build physical solutions to real-world problems. But I was facing my own real-world problem to solve: As I walked the Addis Ababa Merkato, the largest open-air market on the continent of Africa, I failed to see any wood or plastic that matched my mental model. Where I expected smooth planks of wood and uniform sheets of plastic, all I saw were slender trunks of eucalyptus and empty plastic jerry cans that used to hold palm oil. It seemed like none of this could be used as building materials, until I saw a road divider made of cut-up pieces of the yellow and blue plastic and bent wire. Seeing this solution produced an epiphany: culture and context is a lens through which we pre-assign rules about building materials but actually, anything is a raw material.

Most of the world builds with these types of readily available materials, prioritizing functional solutions over aesthetics. This often results in rapidly produced solutions that don't look pretty but work just fine. Prototypes of this sort often are composed of pieces taken out of their original context, found objects, or a repurposing of functions. We call this method of prototyping *low-fidelity prototyping* because it is low-cost, quick, and often one-off.

In the Oshman Engineering Design Kitchen at Rice University we have a makerspace with all of the conveniences one would expect: laser cutters, 3D printers, machining capabilities, electronics capabilities, high-quality machine components. Despite all this, we champion the use of readily available materials and practical ingenuity to solve problems before turning to the tools and materials of our well-stocked makerspace. This is reinforced by the presence of our low-fidelity prototyping cart, full of items that students can combine quickly to solve problems: felt, foam, glue, scissors, tape, K'nex, styrofoam balls, and paper. Those who employ low-fidelity prototyping throughout the life of a project end up producing more prototypes, and these final prototypes are often more refined than the prototypes produced by those who immediately try to jump into making with advanced manufacturing.

With much of the world at a standstill and sheltering at home, many of us are facing limited resources to solve our and the planet's real-world problems. Low-fidelity prototyping is a method everyone can employ in their own home by building a Low-Fidelity Prototyping Kit. This kit is like a first aid kit but for treating problems, not injuries. These materials are low cost and can be combined in a variety of ways. One tip: The best application of low-fi prototyping involves focusing on the function of an object, breaking the cultural and contextual lens you are used to. You might assemble a kit with completely different items; this is OK! We've organized materials by function and based on what most North Americans might have in their immediate surroundings. Don't worry about organization of your kit, just put everything in a plastic bag, another raw material for your building! For starters:

- **Something elastic:** rubber bands, bungee cords
- **Something that covers large areas but bends:** aluminum foil, saran wrap
- **Something planar that spans space:** cardboard (an unrecognized natural resource), paper plates
- **Something that holds things together:** string, binder clips, clothespins, wire
- **Something that joins 2D or 3D materials:** white glue, super glue, tape
- **Volume filling but inert:** styrofoam, balled-up paper
- **Something rigid:** popsicle sticks
- **Cuts objects:** scissors, utility knife
- **Something that can change shape or be molded easily:** clay, play dough ❷

COOPED UP WITH THE KIDS

Keep everyone engaged, entertained, and educated with Family Maker Camp, live now!

Written by Mario the Maker Magician

Family Maker Camp is a virtual opportunity to keep creating, making, and exploring ways to learn new things as a family. *Make:* has gathered the best teachers, makers, projects, and tutorials and created daily live programming available for free via social media. It's accessible, it's live, it's interactive, and it's supporting families, maker style!

My name is Mario Marchese, otherwise known as "Mario the Maker Magician," and I am the official emcee of Family Maker Camp. I help keep families in the loop about what's happening each week, and every Friday I get to host a punk rock dance party and DIY magic session, where kids from all over the world join together to have fun, be silly, make some magic, and see some of the robotics from my theater show.

Because of the school and business closures and sheltering-in-place that Covid-19 has forced, we all need connection right now. Maker Camp jumped in early to help forge an online community of encouragement and support. The real-time, live, interactive format is vital to our well-being! My favorite sessions have been the light-hearted ones, where I interact with viewers. I'm taking the opportunity to do things I don't do in my theater show, like revealing the inner workings of some of my sacred robots and teaching my favorite magic lessons that kids can do with items in their homes, and I'm making up songs with kids' comments and getting everyone jumping up and down!

Beyond that, Maker Camp helps us remember that the traditional path of education is not the only way. I hope that despite this trying time, Maker Camp can help some kids and families realize their own amazing potentials.

FOR THE GROWNUPS

I'm not just the Maker Camp host, I'm a parent too. If I have to sew something for a few hours, or solder something, my kids are always watching and wanting to help, and I find myself with the conundrum all parents face: how to let them be a part of what I do while maintaining some sense of efficiency! But I've learned that the extra time spent is worth it when we see our 8-year-old daughter sewing doll clothes from old T-shirts, or when we see our 5-year-old designing his own

OUR GREAT GUESTS

Maker Camp brings fun shows throughout each week. Check out these incredible participants from this season and join in to watch their segments:

- CodeJoy (codejoyedu.com)
- Dr. Sparks (doctorsparksshow.com)
- Kathy Ceceri (kathyceceri.com)
- John Collins (thepaperairplaneguy.com)
- Brown Dog Gadgets (browndoggadgets.com)
- Kaleidoscope Enrichment (kaleidoscopeenrichment.com)
- STEM Labs (stemlabs.ca)
- Stupid Robot Fighting League (stupidrobotfighting.com)

Katie Rosa Marchese

toys on Tinkercad. Our kids have been raised on the road, with the resourcefulness that comes along with that, and that's something we're super proud of.

Old patterns may be impossible now. Let go of what you can't hold onto and try something different. More than anything, be gentle with yourselves! Even sitting on the couch and watching movies with your kids is educational. The other day, my dad said that the sky seems to be a brighter blue than he remembers. Maybe less pollution? Or maybe we never took the time to look up?

I look forward to seeing you at the next Family Maker Camp. You can build some of our fun and easy projects on the following pages, and tune in at makercamp.com. ✺

MORE ABOUT MARIO

As a professional performer, I normally tour with my family theater show, together with my wife, Katie, and our two kids, Gigi and Bear. The show is an interactive performance full of DIY electronics mixed with homemade magic. Kids see a show full of props made from cardboard, 3D printed elements, and robots made from recyclables, and hopefully realize that nothing I do is out of their own reach. See more at mariothemagician.com.

WORLD WIDE BOOMERANG

This easy-to-build flyer comes back, indoors and out
Written by Slater Harrison

TIME REQUIRED:
30 Minutes

DIFFICULTY:
Easy

COST:
Free if you use a recycled file folder and beg some tongue depressors from the school nurse; or pennies if you buy them, but you have to buy a bunch.

MATERIALS
» **Wood tongue depressors, 6"/155mm long (3)** Also called "senior" tongue depressors or sold as "craft sticks." These are the wide type. Shorter, skinnier popsicle sticks won't work.
» **File folder or thick cardstock** One folder makes two boomerangs.

TOOLS
» **Glue** Super glue or hot glue gun and glue stick work well.
» **Scissors**
» **Tape, clear or masking**
» **Marker**
» **Printed template** Download the PDF at makezine.com/go/world-wide-boomerang and print it at 100 percent (don't "fit to page").

SLATER HARRISON
was an engineer in Bangladesh and a technology teacher in Pennsylvania. He and his wife Naomi work the sciencetoymaker website and YouTube channel.

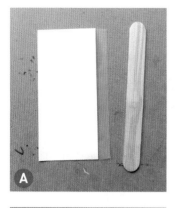

A

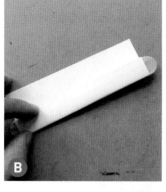

B

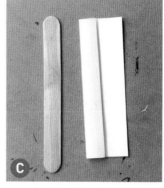

C

D

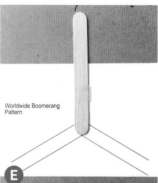

Worldwide Boomerang Pattern

E

Worldwide Boomerang Pattern

F

Mike Senese

When I asked boomerang ninja Logan Broadbent to visit the local high school where I volunteer, he wowed us with his signature stunts that have fueled his videos to tens of millions of views, but he spent most of his time in our design and manufacturing class showing us the science that makes a boomerang work and offering strategies to develop our own designs.

After lots of iterations and some evolutionary dead-ends, we proudly present our World Wide Boomerang. I've made it successfully with middle school and high school classes, and it fits into one class period. It's safe (doesn't hurt people or break windows), costs pennies, flies in a small yard, is adjustable for advanced flying, and yes, really comes back. It's an intuitive way to experience why the airfoil shape is so essential to aviation.

1. MAKE THE AIRFOIL SLEEVES

The airfoils are made from file folder material that we'll fold and tape to form sleeves, which slide onto the wooden spars. Cut a 5⅛"×2⅜" rectangle from the file folder (or cut out and tape the template to the folder as a guide).

Tape the long edge of the card to the long edge of a tongue depressor (Figure **A**). With the taped side facing down, fold the paper over once at the tape, then fold a second time at the paper and make a hard crease (Figure **B**). Remove the stick from the tape, but leave the tape on the paper (Figure **C**).

Reposition the stick into the valley of the second fold. Fold and tape the edge of the paper over the tongue depressor and onto the other part of the paper to make a sleeve that easily slides off the stick — and it's a perfect airfoil! Make two more and set them aside (Figure **D**).

2. GLUE THE SPARS

Tape the first tongue depressor spar to the pattern (Figure **E**). Apply hot glue to the hub and quickly line up another spar with the pattern lines and overlapping the first, then push down before the glue cools.

Do it again for the third spar (Figure **F**). The spars give the boomerang its strength.

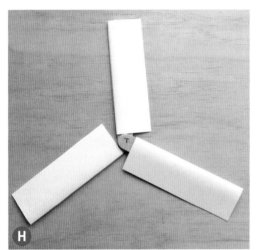

G

H

3. ADD DIHEDRAL

Our boomerang needs a little bit of *dihedral* — that is, an upward slope of the wings. In the case of our boomerang the dihedral is subtle enough that you can't necessarily see it, but you *can* test for it. Spin the assembly on a flat surface, then flip it over and spin again. Choose the side that spins more easily and mark the top of the hub with a "T" (Figure **G**). If both sides spin about the same, choose a side and gently bend the spars up into a curl until it spins freely. The side with the T will always face toward you when you throw.

4. RIGHT- OR LEFT-HANDED?

For right-handed throwers, start with the T faceup and a spar pointing toward you. Slide an airfoil onto the spar so the tape seam faces down and the hole for the stick is on your right side (Figure **H**). For left-handed throwers, the hole goes on the left side. Repeat for the other spars.

Tape them in place and you're ready to fly!

THROW IT!

Boomerangs generate *lift* like helicopters, but by throwing them almost vertically, much of the lift is sideways. That's what makes them come back (along with some *gyroscopic precession* that curves the flight path, and also flattens out the boomerang at the end for a gentle hover down into your hands).

- Remember that the T faces toward you when you throw.
- Hold a wingtip and throw it — with lots of spin — angled at about 1 o'clock if you're right-handed or 11 o'clock for lefties.
- Throw it straight ahead. Boomerangs have their own lift, so you do not need to throw it upward.
- Bend the spars for more dihedral to fly higher and hover down gently. If it's not coming back, gently twisting the wings so the blunt end is higher (aka increasing the *angle of attack*) will increase lift and make it turn back. ⊘

Download the template pattern and watch step-by-step video instructions at makezine.com/go/world-wide-boomerang.

1+2+3 LED THROWIES

A classic! Toss a bunch of these inexpensive little lights to add color to any ferromagnetic surface in your neighborhood   Written by Graffiti Research Lab

YOU WILL NEED:

- » LEDs, 10mm, diffused, any color(s)
- » Batteries, CR2032 3V lithium
- » Magnets, NdFeB disc, Ni-Cu-Ni plated, ½"×1"
- » Strapping tape, 1" One roll will make many.
- » Epoxy, conductive (optional) Weather-resistant alternative to tape.

1. TEST THE LED

Pinch the LED's leads to the battery terminals, with the longer lead (the anode) touching the positive terminal (+) of the battery, and the shorter lead (the cathode) touching negative (−). Confirm that the LED lights up.

2. TAPE THE LED TO THE BATTERY

Tape the LED leads to the battery by cutting off a 7" piece of strapping tape and wrapping it once around both sides of the battery. Keep the tape very tight as you wrap. The LED should not flicker.

3. TAPE THE MAGNET TO THE BATTERY

Place the magnet on the positive terminal of the battery, and continue to wrap the tape tightly until it's all done. The magnet should hold firmly to the battery. That's it — you're ready to throw (or make a few dozen more). Throw it up high and in quantity to impress your friends and city officials. ⊘

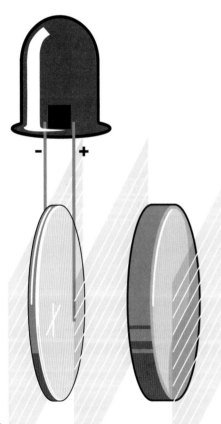

−   **+**

X

NOTE: The battery's positive contact surface extends around the sides of the battery. Don't let the LED's cathode touch the positive terminal, or you'll short the circuit.

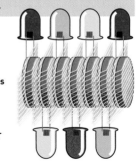

- Throwies naturally chain together in your pocket, making multi-segmented *throwie bugs*, which will also stick to metal surfaces if they aren't too long.

- Your throwies will shine for about 1–2 weeks, depending on the weather and the LED color.

GRAFFITI RESEARCH LAB (graffitiresearchlab. com) is dedicated to outfitting graffiti artists with open source technologies for urban communication.

Kirk von Rohr

1+2+3 IMAGE DUPLICATOR

Build your own mechanical device to replicate images with simple items found at home

Written by Cy Tymony

YOU WILL NEED:

- Thick white cardboard
- Pencils (2)
- Paper
- Paper clips (4)
- Paper clip boxes (2)
- AA battery or other small weight
- Marking pen
- Transparent tape
- Scissors

1

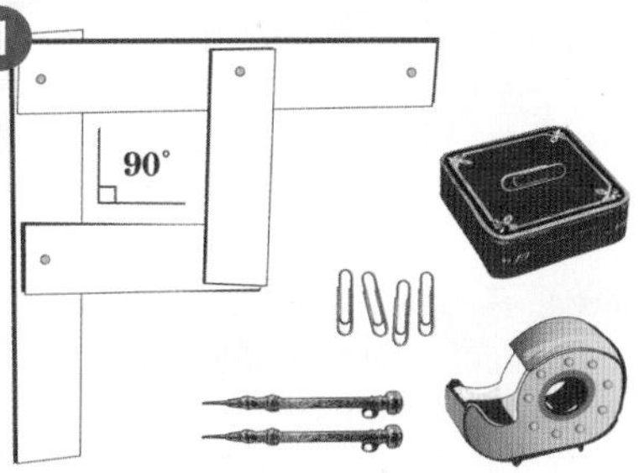

2

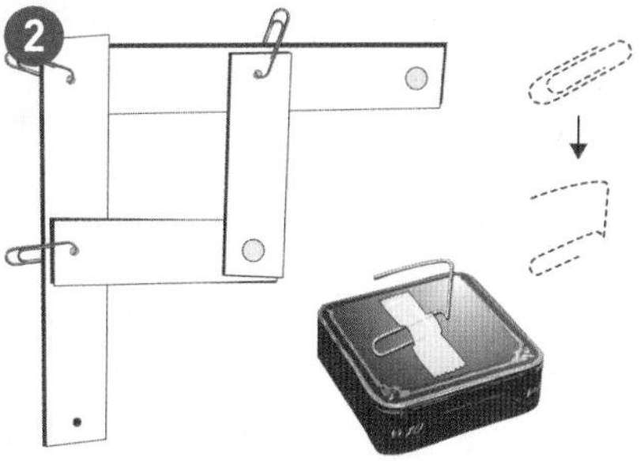

3

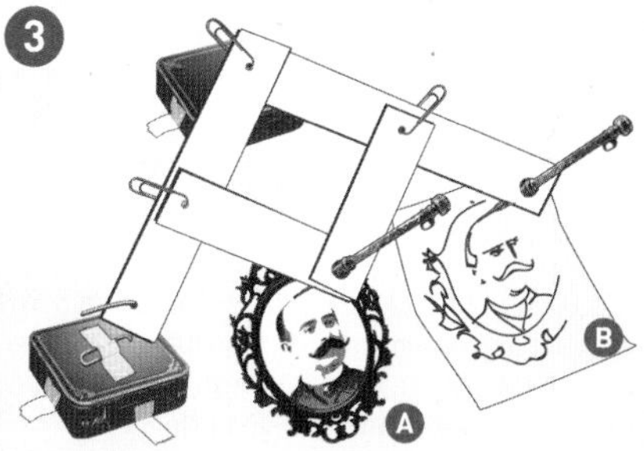

Alison Kendall

With a few everyday items, you can make a *pantograph*, an image duplicator that allows you to use one pencil to trace an image while another pencil follows its path in parallel to draw a copy!

1. CUT OUT AND POSITION STRIPS

Cut 2 cardboard strips measuring 2"×4", and 2 more 2"×8". Place them at right angles, with the smaller pair on top of the larger (Figure 1).

2. LINK STRIPS WITH PAPER CLIPS

Cut 4 tiny holes in the strips and slip paper clips into 3 of them, as shown in Figure 2. Bend up the end of another paper clip, as shown, and tape it to the top of a paper clip box.

3. SECURE TO TABLE

Cut 2 holes in the strips just large enough for 2 pencils to fit snugly and stand erect, as shown in Figure 3. Slip the tiny hole at the end of the left-hand strip over the paper clip that's taped to the top of the box, and tape the box to the table.

Place a second paper clip box under your image duplicator where the 2 large strips meet, to keep it level. To ensure that pencil A presses against the paper properly, you can add weight by taping a AA battery underneath the strip.

USE IT.

Place the original image under pencil A, and a blank sheet of paper under pencil B. Trace the original design with pencil A. Pencil B will follow along, drawing the image on the paper.

Experiment with different lengths of strips to make larger and smaller copies of the design.

CY TYMONY is the author of the *Sneaky Uses for Everyday Things* book series. He lives in Los Angeles. sneakyuses.com

1+2+3 SMARTPHONE PROJECTOR

Show off your mobile photos and your phone hack savvy by turning your smartphone into an inexpensive projector Written by Danny Osterweil and Photojojo

YOU WILL NEED:
- » Shoebox
- » Paper clip
- » Smartphone
- » Magnifying glass with low magnifying power
- » Pencil
- » X-Acto knife
- » Electrical or black duct tape
- » Matte black spray paint or black paper (optional)

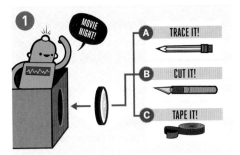

PAPER CLIP ORIGAMI!

Slide projectors are great, but outdated, and digital projectors cost a bundle. Fortunately, you can make your own.

1. PREPARE THE PROJECTOR BOX
For best image quality, paint the inside of your shoe box black or tape up some black paper. On a short side of the box, trace the outer edge of your magnifying glass and cut it out. Cut a small hole at the back of your box for your phone's power cord. Tape the magnifying lens in place, and make sure there are no holes to let light in.

2. MAKE A STAND AND FLIP IT
Bend your paperclip into a cellphone stand. When light passes though the lens, it gets flipped, so the picture from your projector will come out upside-down. Visit makezine.com/projects/5-smartphone-projector for instructions on how to achieve a screen flip.

3. FIND YOUR FOCUS
Project onto a bare white wall. Position your phone in its stand near the back of the box and walk the box forward or backward until the image starts to come into focus. Fine-tune the focus by moving the phone forward or backward in the box. Set your phone's photo app to slideshow mode for a hands-free experience. For best viewing, turn the screen brightness of your phone all the way up, put on the box top, cover any windows, and turn the room lights down.

Thanks to Instructables user MattBothell for inspiring this project! ⊘

PHOTOJOJO (2006–2018) was a majestic unicorn of an online community and photography gear store. Today the team carries on as purveyors of photo books and prints at Parabo Press. parabo.press

Julie West

1+2+3 CUSTOM MEMORY GAME

Give an old standby some new flavor and create a perfect rainy-day activity Written and photographed by Julie A. Finn

YOU WILL NEED:
» **Paper** plain, colored, or printed, for making matching image pairs
» **Scissors**
» **Glue stick**
» **Laminating sheets or a laminating machine**
» **Fancy paper** such as scrapbooking or wrapping paper, for the faceup and facedown sides of your memory cards

Memory is the perfect game. A kid can play it solo or in a group, it can be easy or challenging, and it's educational to boot.

1. PREPARE YOUR IMAGE PAIRS
Cut out matching image pairs from paper, such as numbers, colors, or words. (Here I used paint swatches.) Trim them all to about the same size.

2. CUT YOUR CARDS
Find a template for your memory cards — a cassette tape box, coffee mug, etc. — and cut out identical faceup sides from your fancy paper for each card in your game.

Also cut out a facedown side for each card — this can be from the same or different paper as your faceup side. These don't have to be identical, because they won't be facing up at the start of the game.

3. GLUE AND LAMINATE
With a glue stick, lightly stick your faceup and facedown sides together, then stick one memory game image on each facedown side.

Laminate your cards for durability, or take them to a copy shop to be laminated.

USE IT.
While simple identical color matches work well for a memory game, consider tweaking your game toward a specific educational experience or a specific kid. Try matching colors with color words, for instance, or pictures with Spanish words, or math equations. Keep your kids guessing! ✪

JULIE A. FINN blogs about all the wacky hijinks involved in parenting and the crafty life at craftknife. blogspot.com.

Digital Revolution

Written by Justine Haupt

Take a break from your touchscreen and dial up your friends the traditional way — with a real dial

When touchscreens first came on the scene I was, like everyone, excited for the future of user interfaces. What wound up happening is horrible.

Instead of getting the exacting control depicted in science fiction, we received abstruse devices that don't make it clear whether a program ("app") is open or closed, which force us into a world of childishly colorful windows that give no quarter in the battle for precision, and which are used to maintain a culture of hyperconnectivity.

It felt like the whole world went from using emails and telephones to *expecting* texting in the blink of an eye, and I found myself hardened to reject it all. I stuck with a flip phone, and with the exception of a month-long experiment with a smartphone, have been happy to have only basic phone functionality and a good excuse not to text, because everyone knows how awkward texting would be on a flip phone. But even the flip phone isn't immune to stupid user interface design.

I'd been itching for years to use a rotary dial as a data entry interface for some project. They're easy to interface with electrically, so I imagined a rotary numpad replacement for a computer, or perhaps as a means to enter sky coordinates on a steampunk telescope mount. But when I noticed the Adafruit Fona, the possibility of building a cell phone seemed almost easy, and adding a rotary dial seemed just quirky enough to be fun. I could finally have a phone that would behave the way I want it to. It could be entirely mine and as tactile as I could imagine, and I could just *not* write the firmware (Arduino sketch) needed to allow texting. The point wouldn't be to have something wacky, but to have to be a real, usable phone that I would want to use as my primary phone; a

TIME REQUIRED: A Weekend
DIFFICULTY: Difficult
COST: $300

MATERIALS
» **Rotary cellphone main circuit board** Make it from the KiCad design files or send them to a PCB maker like pcbway.com. I've also made it available as a kit from skysedge.us.
» **Rotary dial, Western Electric model 10A**
» **Adafruit Fona 3G cellular breakout board** American or European version, depending where you live, Adafruit #2687 or 2691
» **Compatible 3G SIM card** I use AT&T prepaid.
» **Adafruit eInk Breakout Friend** #4224
» **ePaper display, 2.13" flexible** Good Display #GDEW0213I5F, Adafruit #4243
» **Battery, 1.2Ah LiPo with JST connector** such as Adafruit #258
» **Antenna, quad-band GSM with SMA connector** Adafruit #1858
» **SMA bulkhead connector w/µFL pigtail** Adafruit #851
» **3D printed enclosure and buttons**
» **Labels**
» **Clear plastic packaging material** to be cut into "windows" for enclosure
» **Threaded inserts, M2, heat-set** McMaster-Carr #94180A307
» **Machine screws, M2, socket head: 6mm (2), 10mm (3), and 14mm (1)**
» **3M 468P transfer adhesive**
» **Tape, Scotch double-sided**
» **Epoxy or plastic model glue**

TOOLS
» **Soldering station with fine chisel tip**
» **Solder, fine-gauge** e.g. 0.20"
» **Solder flux**
» **Tweezers, fine**
» **Tweezers, coarse** or small needlenose pliers
» **Hex keys** aka Allen wrenches
» **Scissors**
» **Hot glue gun**

JUSTINE HAUPT is a developer of astronomy instrumentation at Brookhaven National Laboratory. She was a primary contributor to sensor development for the Vera Rubin Telescope / Legacy Survey of Space and Time, for which she designed electro-optical test stands. She is currently working on the Baryon Mapping Experiment (a 4-dish, 21cm cosmology demonstrator telescope) and a quantum networking project. Her competencies span optical, mechanical, and RF engineering.

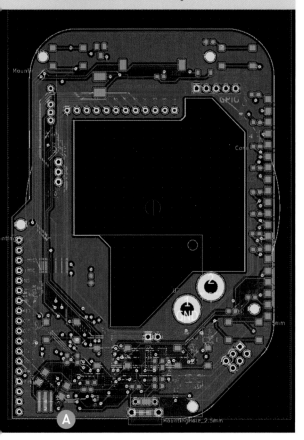

A

demonstration of a phone that goes as far from having a touchscreen as possible while being a "slap in the face" to smartphones, because in a few ways it would actually be *more* functional.

It would have a real external antenna with superior reception. It would have dedicated physical keys for the one or two people I call most, and voicemail. It would have instantaneous signal strength and battery metering with more resolution than just four or five bars. It would have an e-paper display — an entirely underused technology that's bistatic, meaning it doesn't take energy to display a fixed message and is absolutely beautiful. The power switch would be an actual slide switch that doesn't require holding down a stupid button that makes you guess about what it's really doing. But most importantly, I would have control. If I wanted to change something about the phone's behavior I would just do it, and if there was something I don't like there would be no one to blame but myself.

The finished phone fits in a pocket. It's reasonably compact. Calling the people I most often call is *faster* than with my old phone, and the battery lasts more than a day. It is now, in fact, my primary phone.

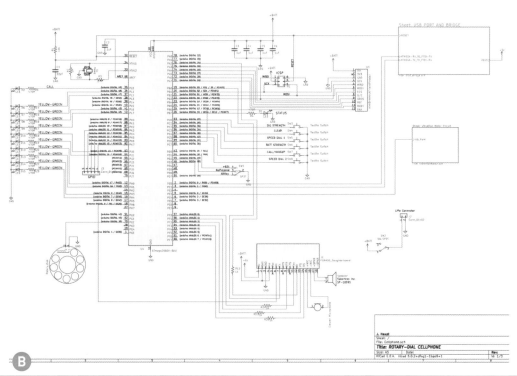

B

All my design files and detailed build steps are available at skysedge.us/unsmartphones/rotarycellphone_3g/build. Here's an overview of putting one together.

ASSEMBLE THE PCB

The heart of the phone is a custom 4-layer PCB. I made both the electrical design and board layout in KiCad, a fantastic open source electrical design suite that puts some commercial EDA software to shame (Figures A and B).

I sent the design files to a PCB "board house" for manufacture. With the company I use, a batch of 5 PCBs comes out to about $60 delivered. I did all the soldering by hand. Most of the components are surface-mounted, which might scare people, but with a good soldering station, tweezers, and some patience, this is a doable hand-solder job. As it's a double-sided board, expect to spend a long afternoon working on it. You'll also need to solder on an Adafruit ePaper Breakout Friend, which provides control for the e-paper display, and solder a header to the Adafruit Fona 3G, which is our cell radio daughterboard (Figure C).

UPLOAD THE SOFTWARE

With the board together you can flash the firmware (i.e. upload the Arduino sketch). This will require a few extra libraries loaded into the Arduino IDE, like Jean-Marc Zingg's GxEPD library for controlling the e-paper display, a Fona library from Adafruit, a couple other libraries, and the MegaCore board support add-on from MCUdude which will allow the Arduino IDE to work with the ATmega2560V microcontroller.

I always use an in-system programmer when uploading Arduino sketches. Instead of dealing with the USB port, which can be flaky for this purpose, an AVR-ISP-Mk2 programmer can be used instead and this always *just works* for uploading firmware. They go for about $20 at Digi-Key and Mouser and plug into the ICSP header which is always present on Arduino-like boards (in case you didn't know what it was for), and then, to initiate the upload we just press Ctrl+Shift+U within the Arduino IDE. No need to select the USB device assignment. The board settings within the Arduino IDE should be as follows:

Justine Haupt

Board: ATmega2560, Clock: External 8MHz, BOD: BOD 2.7v, Compiler LTO: LTO enabled, Pinout: Arduino MEGA pinout, Bootloader: Yes (UART0).

BUILD THE MECHANICALS AND ENCLOSURE

The specific rotary dial needed is a Western Electric model 10A (Figure D). This model comes in many sub-variants, but as long as it's some kind of 10A it should work. These can be found as part of certain Trimline rotary telephones (check eBay) or from specialty suppliers of old phone parts. I have more notes at skysedge.us/unsmartphones/rotarycellphone_3g/build/dials.

You'll need to modify the dial to fit the PCB and 3D printed enclosure properly, which involves hacking off the metal mounting tabs around the dial's perimeter, filing, and cutting down the potted mounting feature for the dial's reed switch. Depending on the specific variant, it may also be necessary to remove the electrical leads for the reed switch in the process of cutting it, and then

Justine Haupt

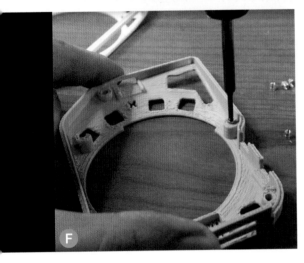

solder them back on afterwards. Once the dial passes a test-fit with the PCB and the enclosure (Figure **E**), screw the dial onto the PCB and check that using it activates the LEDs on the board when the board is energized.

3D print the enclosure and buttons; the STL files are available on my site. Press the M2 heat-set threaded inserts into place using a soldering iron (Figure **F**).

The SMA to µFL adapter cable can be installed into the backside of the enclosure, and as an optional step you can add clear plastic "windows" to the cutouts in the enclosure which provide visibility to certain parts of the board while also protecting against lint ingress. I use product packaging from the garbage for this.

FLEXIBLE E-PAPER

Next comes the trickiest part of the whole process: bonding the flexible e-paper display. While marketed as flexible displays, these are fragile. I killed about six of them in the process of understanding their limitations and figuring out how best to bond them to the enclosure.

The flex cable is resilient and can be folded on itself, but the section between the flex cable and the white portion of the display is extremely delicate, as it contains an encapsulated integrated circuit which is *not* strain relieved. You can actually see the silicon below the top surface of the kapton. Bending it in the "wrong" direction of the display will snap it in two. Bending in the "correct" direction will disrupt whatever bonding is done between silicon and the traces in the kapton flex material and cause display artifacts or complete failure. So keep this portion as flat as possible; I've found that the best thing is to epoxy it to a flat segment of the enclosure, or perhaps to use transfer adhesive (Figures **G** and **H**).

Much effort also went into designing the bend around the back of the display to be as tight as it can be without causing display glitches, and it's not clear that I've gone far enough in this regard.

CLOSE IT UP

The buttons can now be set into the enclosure and the main board lowered into place. Before doing this, you want to have a compatible SIM card installed in the Fona (AT&T prepaid works for

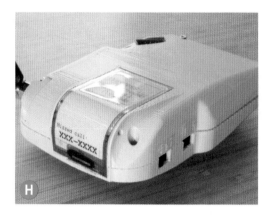

me in the U.S.), and of course the Fona should be inserted into the socket on the main board. Some care should be taken to be sure cables for the LiPo battery and the SMA connector are routed well and that the battery is nestled next to the Fona so that the back of the enclosure will fit, and there's a tricky process involved in mating the FPC connector to the eInk Friend daughterboard.

With the main board installed and the rotary dial sticking out the other side of the enclosure, we can install the back of the enclosure and finish bonding the e-paper display in place, this time using simple double-sided tape, which permits

you to remove the display again without damaging it if you need to open the phone for maintenance.

Assuming the recommended SMA duck-style antenna is used, after attaching the antenna another small piece gets screwed on over the nut-section of the SMA connector. I call this the "SMA plug" and it retains the SMA nut to be fixed in a slightly loosened position that allows the antenna to swivel, even though it isn't designed for articulation. The joint on an actual articulating antenna takes up enough space that I felt the need to work out this hack on a more compact "fixed" SMA antenna, and it works quite well.

The only remaining detail would be to decorate it with labels for the buttons and switches, as desired. I use an Epson LabelWorks label maker with either black or white ink on clear backing, depending on what color casing I've 3D printed. The labels need to be cut down to size with scissors, and to get them to stick well to the ridges/steps in the 3D printed plastic it's helpful to put down a small blob of glue to fill the gaps arising from the steps from the 3D printing. Plastic model glue has worked well but there are likely even better options.

A working rotary-dial cell phone should now be sitting before you. Cool. ◉

My Monkey Companion Bot

Written by Jorvon Moss

TIME REQUIRED:
2 Weekends

DIFFICULTY:
Intermediate

COST:
$60–$80

MATERIALS

» **Chemion customizable LED glasses** Amazon #B01B41PHJM
» **Micro servomotors, SG90 (5)**
» **Arduino-compatible microcontroller, 5V, 16MHz** I used the Adafruit Pro Trinket 5V, Adafruit #2000, but you can use the newer ItsyBitsy or other small Arduinos.
» **Boost converter, 3.7V to 5.2V DC** Adafruit PowerBoost 1000C, Adafruit #1944
» **Battery, LiPo, 3.7V, 2000mAh**
» **Battery charging/management board** such as the Adafruit LiIon/LiPoly Backpack, #2124
» **Micro switches (2–4)** Right now I'm using one for servos and one for lights. There's room for 2 more, to toggle whichever systems you add.
» **3D printed body parts** Download the free files for printing at makezine.com/go/dexter.
» **Screws (many)**
» **Acrylic sheet, ⅛" thick, about 4"×8" (optional)** to make the eye covers
» **LEDs (optional)** I used Adafruit NeoPixel rings.
» **Copper tape (optional)** I used it as a touch sensor.

TOOLS

» **Computer with Arduino IDE software** free at arduino.cc/downloads
» **Hot glue gun**
» **Drill**
» **Screwdrivers**
» **Soldering iron and solder**
» **Solder sucker or desoldering braid** to remove the Chemion LCD display
» **3D printer (optional)** You can print the files yourself, or send them to a printing service.

JORVON "ODD JAYY" MOSS is a tinkerer, creator, designer, and maker of gadgets and robots in Compton, California. He's also a big fan of steampunk and cyberpunk. @Odd_Jayy

Build a fun robot monkey you can customize, wear with you on adventures, or use as an avatar

Allow me to tell you the tale of my robot monkey. Last year was a big year for me and my companion bots, entering the eye of the maker world: I even got invited to talk at Hackaday Superconference. Thanks to that event, and me being told about it three months prior, I began working on my robot monkey companion, Dexter.

What's a companion robot? To me, it's a robot you can take with you on any type of adventure, the way R2-D2 rolls next to Luke Skywalker. I started making them in 2018 and couldn't stop. When adults see me with a companion bot they might think it's weird or ask me a polite question, but kids get so excited: "OMG, it's a robot!" They know where it's at.

Dexter was a 4-month journey of failure. Dexter V1 was fragile like a baby; I took him out one day and he snapped in half, wires everywhere. Dexter V2 was much sturdier, and I added fingers and constantly kept upgrading him; a lot of things worked, and did not work. I even upgraded him to V3 at the Hackaday event, inspired by some amazing makers there: I gave him his new custom eyes, improved his movements by changing up some screws and adding a power booster, and reprinted a new face for him. Then

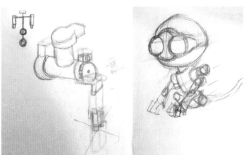

Early sketches of Dexter's design.

Tiffany Chien – bytifanychien.com, Jorvon Moss

I showed him off at Maker Faires in Los Angeles and the Bay Area.

What I'm sharing with you is Dexter V4, the most upgraded version yet. I used Fusion 360, and a bit of SolidWorks, to design him. His eyes are a hacked pair of LED glasses, and I made him with the intention and ability to be upgraded and customized depending on the person making him. He's still not perfect — I drilled my own holes to add screws, and hot-glued a lot of him — but I hope you like him.

ASSEMBLY

I recommend starting with the head, which is made of three parts: the front (face), the back, and the middle plate. I would print these first, then the neck gear and the ribcage. Use these to assemble the head to the neck; that way you'll quickly have a piece of him built.

Next, the servo gears. Print these at high quality; they should fit any micro servo but I know they tend to warp if the quality is low. Add one gear to the neck servo and fit the servo into its place in the rib cage. I recommend you run a servo test and see how the gears mesh and move the neck. I used the Arduino example code at File→Examples→Servo→Sweep. If it works well (or good enough for your specific uses), continue.

The ribcage has two more of the same servo placements, one in each side; add servos there to move the arms, and test those too. In the end I used ball joints to make Dexter's arms because they can be moved in more directions and adapt to his riding position when he's on my back.

I recommend printing the belly next. It's pretty big and will be used to hold most of the circuit inside it. It's got two block-like platforms; use these to attach it to the ribcage using screws or glue.

SERVO CIRCUIT

Now prepare your circuit. Dexter is cool because he can be customized: I made him so that you can edit his systems and make and put your own circuits in him. For his servos I used an Adafruit Trinket Pro microcontroller driven by a single LiPo and an Adafruit voltage booster board (Figure A). With this setup I can move as many as 6 servos at random intervals, so Dexter is always doing something.

Dexter V3 at Downtown L.A. Maker Faire, 2019.

An early Fusion 360 STL.

Head/neck test movement, V1.

Dexter's arm with ball joints

Jacques Alexander Katzoff, Jorvon Moss

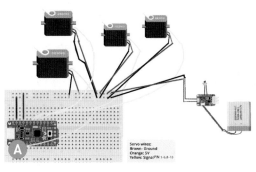

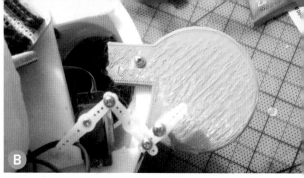

Next the ears. These are pretty simple. Attach a simple micro servo to the back of the inner skull piece as in Figure B.

Attach the servo arms to the ears in such a way that they'll push and pull the ears up and down, then attach each ear with a single screw to the top part of the inner skull, so it can pivot. This gives Dexter his moving ear look (Figure C).

CHEMION EYES

The eyes are next. Dexter V4's "emote" eyes are actually hacked Chemion LED glasses. There are multiple tutorials online on how to hack these and rewire them; my favorite is youtube.com/watch?v=xB0rmM351pg. The eye covers are just laser-cut acrylic and they're hot-glued to the face (Figure D). I control the eyes from the Chemion phone app; previous Dexters had LED matrix eyes controlled by a microcontroller.

Next I recommend printing out the other parts: the butt plate, legs, hands, and feet. It's easy to assemble Dexter, but the print time does take a while. I designed the back of the head plate with four slots for switches, so you can wire up the servos, LEDs, etc., separately with the ability to turn them on and off. I usually put the circuit in Dexter's stomach, so some of the wires show through his neck (Figure E); I usually hide these with a scarf.

Oh, also his tail. Almost forgot, I used these modular cable covers I found on Thingiverse (Figure F) designed by Greg "sirmakesalot" McRoberts in Toronto (thingiverse.com/thing:25872) to make his tail, and used armature wire as a base so you can bend his tail where you want it, and you won't have to worry too much about his tail hitting people.

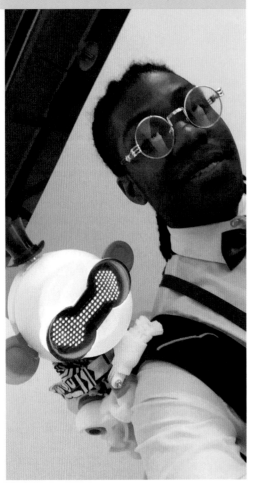

Now for Some Monkey Business

MONKEY ON YOUR BACK

To wear Dexter, I use a GoPro mount to attach his chest plate to a backpack strap, so his center of gravity stays low, behind my shoulder. An earlier version sat up high on my shoulder but that was too tippy (Figure **G**).

YOUR BASIC RANDOM MOVEMENT SERVO CODE

This Arduino code moves the servos 0°–180° in random intervals; in my opinion, this just helps simulate life. You can rewrite or replace it with your own Arduino code to make Dexter move however you want. There are 5 servos in Dexter currently but the code provides for 7 because I wanted Dexter to have the ability to be hacked or updated later, by me or other people. You can find the code on my Dexter project page at hackster. io/Odd_Jayy.

EXPANDABLE AND UPGRADABLE

Lastly I wanted to add what's special about Dexter. The reason there are 4 switch holes in the back of the head design is so you can add other circuits to him (Figure **H**).

In different versions of Dexter I have added an LED circuit to glow inside (NeoPixel rings) and a touch sensor to make him do high fives, hahaha (copper tape on his forehead, connected to an Adafruit Circuit Playground microcontroller inside, Figure **I**). There's room for whatever sensors, brains, and behaviors you want to add.

THE FUTURE

I have other, crazier ideas for companions, like a dragon with a real flamethrower — not sure I'll be able to bring that one to work — a fairy robot, and a fox I've already started sketching. Right now I'm working on a smaller dragon that blows smoke from a vape pen (see sidebar). I'll still work on a bigger dragon one day, but I want to continue upgrading Dexter and coming up with more ways to make more fun bots. *Star Wars* doesn't have to have all the cool droids, we can make some cool ones too! ◘

G

Jorvon Moss

My Companion Bot Menagerie

I'm a workaholic. I love building robots. It's been a busy year! You can find all my companion bots at hackster.io/Odd_Jayy.

Asi
Easy-wear arachnid with LED matrix eye to scare neighbors. Inspired by African spider god Anansi, the trickster/storyteller.

Odin
Small wearable bird is a quick build for kids. Inspired by Alex Glow's familiar ("Archimedes: AI Robot Owl", *Make:* Volume 66).

Sun
Astronaut-themed wearable bot, inspired by Sophy Wong's amazing spacesuit ("Cosmic Cosplay," *Make:* Volume 69).

Prometheus
Alien-like design with analog, animatonic moving eyeballs instead of LED matrix animation.

Widget
Skyrim-inspired dragon with optional vape-pen smoke module. Built with Adafruit Crickit and Make: Code.

Backyard Wind Power

Written and photographed by Ulrich Schmerold, Translation by Nig Oltman

Build a simple generator to power your small projects from the breeze.

For low-power applications around your home and yard, professional wind turbine installations are just too pricey. If all you need is a bit of juice for LED illumination or a Raspberry Pi Zero project, paying thousands for a small wind energy system seems disproportionate. And for experiments in school, the cost and time required should be minimal, too — schools are often strapped for cash. In this article, we'll show you how to build your own small wind-power installation from old bike parts and stuff from the hardware store. With just a little more than a breeze, it can provide about 1 watt of power. That's enough to charge a small battery, so you'll still have power when it's calm.

This small wind turbine is more of an experiment to teach you the basics; it won't provide you with 100 percent reliable power. No miracles here! Also, please beware of strong winds and storms: this machine is not designed to handle that kind of weather and would likely disintegrate. You must protect it from such potential damage, as flying debris could cause injury.

In contrast to the typical three-blade commercial wind turbines, we use a vertical rotor shaft. This eliminates the need for wind directional tracking and leaves us with a very simple design. Essentially, it's just a vertically mounted bicycle wheel with a hub dynamo. For rotor blades, we use eight "half-pipes" cut from cheap plastic (PVC) drain pipe, vertically attached to the rim.

Our turbine will start spinning as soon as wind speeds reach about 2 on the Beaufort scale, or 5mph. With a stiff breeze of 20mph or 5 Beaufort (see the conversion table on page 88), it provides around 1W of power output (we measured 147mA at 6.7V).

TIME REQUIRED:
1 Day

DIFFICULTY:
Intermediate

COST:
$80–$150

MATERIALS

» **Bicycle front wheel with dynamo hub, 28"** I bought one new on eBay for €40 but used ones are ubiquitous in Europe. In the USA you can score one on eBay, or buy a cheap Shimano dynamo hub for $50 and install it in an old wheel.

» **PVC drain pipe, 4" or 110mm nominal width, 6' or 2m lengths (2)** I used thin-wall pipe but the exact type doesn't matter.

» **Machine screws, with nuts and large washers (16)** length and diameter depending on your wheel rim

» **Steel water pipe, zinc galvanized, threaded both ends, 1½" diameter** The length (mast height) will be matched to your local conditions.

» **Steel pipe fittings, 1½": end cap (required) and tee (optional)** for water pipe

» **Buck-boost voltage converter, DC–DC** such as Mesa #DSN6009 4A. I recommend 30W output capacity.

» **Electrolytic capacitors, 2200µF (2)** 12V minimum

» **Full-bridge rectifier** 500mA minimum

» **Diode, 1N4007**

» **Heat-shrink tubing or electrical tape**

» **Wire cables and screw eyes (optional)** to guy out a tall mast

» **Concrete, 60lb bag (optional)** to set the mast

TOOLS

» **Handsaw or jigsaw** for cutting thin PVC pipe
» **Drill with bits** for drilling plastic and metal
» **Screwdriver and/or wrench or socket set** to fit your screws/nuts/bolts
» **Soldering iron and solder**

ULRICH SCHMEROLD lives in Bavaria in the south of Germany and builds devices for people with disabilities. He likes to create projects that get people excited about physics.

Ulli also made the Micro Ultrasonic Levitator in *Make:* Volume 65, makezine.com/projects/micro-ultrasonic-levitator, and the Persistence-of-Vision LED Globe in *Make:* Volume 53, makezine.com/projects/persistence-vision-led-globe.

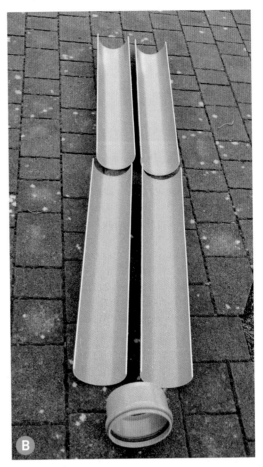

BUILD YOUR BIKE WHEEL WIND TURBINE

Let's start by building the rotor-and-generator unit. You'll be using a mast made from a steel water pipe, probably secured in the ground using poured concrete. Check your local requirements for the foundation and the mast height, and adjust accordingly. Depending on local conditions, you may also need to anchor your mast using wire cables.

1. CUT THE TURBINE BLADES

We used thin-wall PVC drain pipe to make our turbine blades (Figure Ⓐ). In Germany, where we live, this stuff is orange; in North America it's usually white.

Using a jigsaw, you can cut 4 blades from a 6' or 2m length of pipe (Figure Ⓑ). We need 8 blades in total. Take care to cut the pipe exactly in the center — ideally, the blades should all be the same weight.

2. ATTACH BLADES TO GENERATOR

For the generator, we use a bicycle wheel (rim) fitted with a hub dynamo (Figure Ⓒ). Rims made from aluminum work best, as they can be drilled easily. If you're taking parts from a used bike, make sure to remove the tire and inner tube, and any brake discs.

Attach the 8 turbine blades as shown, using 2 screws, nuts, and large washers each, spaced evenly (try counting the spokes), and centered on the rim (Figure Ⓓ).

3. MAKE THE MAST

Make the mast from zinc-plated steel water pipe threaded at both ends (Figure Ⓔ). Drill a 9mm hole into the end cap and tighten your hub nut onto the bike wheel's axle to attach the wheel to the cap (Figure Ⓕ). Once the mast has been mounted securely in the ground (!), you can screw the cap onto the mast.

For erecting the mast, the thread at the

other end may come in handy. You can thread a matching tee piece onto it, and encase the tee in the concrete block that you'll pour in the ground. The concrete should be sufficiently heavy to both support and anchor the turbine, and must be fixed firmly in the ground. Then, whenever a storm comes up, you can simply unscrew the mast from the concrete block and take the turbine somewhere safe.

Don't make the mistake of underestimating the forces created by winds. They grow proportionally to the cube (third power) of the wind speed! If necessary, guy out the mast with wire cables.

4. ASSEMBLE THE ELECTRONICS

Our device is set up for charging a lead-acid battery using the current generated by the dynamo (Figure **G**). The hub dynamo produces alternating current, which we'll convert to pulsating direct current using a full-bridge rectifier. To smooth it out, the pulsating DC is fed into the two 2200μF (microfarad) electrolytic capacitors.

The smoothed DC is then passed to a buck-boost converter (about $10 on eBay) which we'll use as a charging regulator. This will convert any input voltage from 1.25V to 30V into an adjustable, constant output voltage. We'll set the converter's output to be 0.7 volts above the end charging voltage of our battery (compensating for the diode's forward voltage). The 1N4007 diode is required to prevent current flowing back from the battery to the converter.

For example, a 6-volt lead-acid battery has a charging voltage of 7.2 volts. Adding the diode's forward voltage of 0.7V, the converter should be set to an output voltage of 7.9V.

Your electrical load (whatever's going to consume the power, an LED lamp, for instance) will be connected to the battery's output. Be aware that the load must be able to handle the output voltage set for the converter. While the generator itself may only be able to provide a small amount of current, the battery may output several amps. In case of a short circuit, the consequences may be dire (fire hazard). To prevent accidents, you'll need to safeguard accordingly whatever circuit you're connecting to the battery.

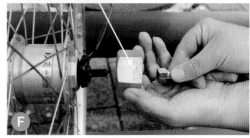

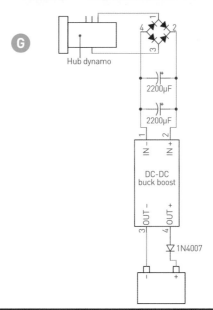

Hub dynamo

2200μF

2200μF

IN –   IN +

DC-DC buck boost

OUT –   OUT +

1N4007

The Beaufort Wind Force Scale

BEAUFORT #	DESCRIPTION	WIND SPEED
0	Calm	< 1 knot, < 1 mph, < 2 km/h, < 0.5 m/s
1	Light air	1–3 knots, 1–3 mph, 2–5 km/h, 0.5–1.5 m/s
2	Light breeze	4–6 knots, 4–7 mph, 6–11 km/h, 1.6–3.3 m/s
3	Gentle breeze	7–10 knots, 8–12 mph, 12–19 km/h, 3.4–5.5 m/s
4	Moderate breeze	11–16 knots, 13–18 mph, 20–28 km/h, 5.5–7.9 m/s
5	Fresh breeze	17–21 knots, 19–24 mph, 29–38 km/h, 8–10.7 m/s
6	Strong breeze	22–27 knots, 25–31 mph, 39–49 km/h, 10.8–13.8 m/s
7	High wind, moderate gale, near gale	28–33 knots, 32–38 mph, 50–61 km/h, 13.9–17.1 m/s
8	Gale, fresh gale	34–40 knots, 39–46 mph, 62–74 km/h, 17.2–20.7 m/s
9	Strong/severe gale	41–47 knots, 47–54 mph, 75–88 km/h, 20.8–24.4 m/s
10	Storm, whole gale	48–55 knots, 55–63 mph, 89–102 km/h, 24.5–28.4 m/s
11	Violent storm	56–63 knots, 64–72 mph, 103–117 km/h, 28.5–32.6 m/s
12	Hurricane force	⩾ 64 knots, ⩾ 73 mph, ⩾ 118 km/h, ⩾ 32.7 m/s

The wind speed scale we use today dates back to the 18th century. Originally, it was made to describe the effects on windmill vanes. British seafarer Sir Francis Beaufort (1774–1857) was by no means the first to publish such a scale; his work descended from that of civil engineer John Smeaton (1759) and geographer/hydrographer Alexander Dalrymple (1790). Even earlier scales were created by astronomer Tycho Brahe (1582), polymath scientist Robert Hooke (1663), and tradesman, rebel, spy, and *Robinson Crusoe* novelist Daniel Defoe (1704). But from 1829 on, Beaufort, who had now been appointed hydrographer to the British Admiralty, shared his scale with all interested parties. The Beaufort Scale has since become a standard.

(Source: Wikipedia, en.wikipedia.org/wiki/Beaufort_scale)

STORM WARNING!

With the electronics assembled, you're ready to wind-power your setup! Enjoy your new capacity as a wind turbine owner.

This wind turbine is meant as an experiment, a low-cost practical demonstration of how wind turbines work in principle, for instance in a school setting. It's not intended to withstand strong gales or severe storms. When not in use, or when wind speeds exceed 6 on the Beaufort scale, it should be dismantled.

The bike wheel and the mounts for the rotor blades aren't designed for permanent operation, in particular with strong winds. We recommend you take your own steps to strengthen this design if you want to make it permanent. (That said, the construction was more stable than expected. I left it in the garden all the time, whatever the weather. Only when a cable tie gave out, the mast fell over and a blade was destroyed.)

Do you operate a wind turbine? We'd love to hear from you at wind.turbine@make.co (send us photos and specs, please). We'll include your contributions in a future report. ◗

MORE WIND POWER PROJECTS:

Otherpower, DIY wind power experts: otherpower.com/otherpower_wind.html

Simple Savonius vertical rotor turbine: macarthurmusic.com/johnkwilson/ MakingasimpleSavoniuswindturbine.htm

Easy PVC wind turbine for schools, from U.S. Department of Energy: energy.gov/eere/education/ downloads/building-basic-pvc-wind-turbine

Mast from the Past

In 2006, New Mexico homesteaders Abe and Josie Connally wrote an excellent how-to in *Make:* Volume 05 for building their Chispito Wind Generator from PVC pipe and an old exercise treadmill motor (makezine.com/projects/wind-generator). Three years later, John Edgar Park built the project on national TV for PBS' *Make: television*; you can follow along with his build at youtu.be/1vnyYWV78zk.

The Chispito is still popular today — Abe and Josie later posted the project on Instructables where it garnered hundreds of comments, and on their own site velacreations.com, where they document all kinds of wonderful off-grid DIY projects. Their solar food dryer, top-bar beehive, and earth block floors were all featured in *Make:* as well (makezine.com/author/abe-connally).

Raised on a mast 10'–30' high, the Chispito will generate 84 watts of power in a 30mph wind; be sure to follow the updated instructions for shaping the blades at velacreations.com/chispito.

WIND GENERATOR BLOW-BY-BLOW

The Chispito Wind Generator is a simple little machine that's great for getting started with wind power. In a 30mph wind, ours gives us about 84 watts, 7 amps at 12 volts.

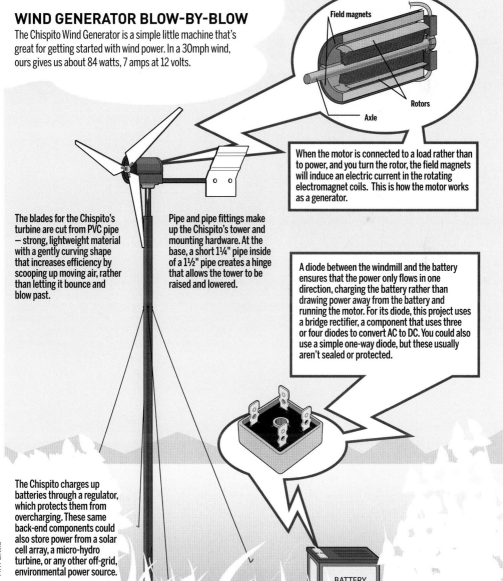

Field magnets

Rotors

Axle

When the motor is connected to a load rather than to power, and you turn the rotor, the field magnets will induce an electric current in the rotating electromagnet coils. This is how the motor works as a generator.

The blades for the Chispito's turbine are cut from PVC pipe — strong, lightweight material with a gently curving shape that increases efficiency by scooping up moving air, rather than letting it bounce and blow past.

Pipe and pipe fittings make up the Chispito's tower and mounting hardware. At the base, a short 1¼" pipe inside of a 1½" pipe creates a hinge that allows the tower to be raised and lowered.

A diode between the windmill and the battery ensures that the power only flows in one direction, charging the battery rather than drawing power away from the battery and running the motor. For its diode, this project uses a bridge rectifier, a component that uses three or four diodes to convert AC to DC. You could also use a simple one-way diode, but these usually aren't sealed or protected.

The Chispito charges up batteries through a regulator, which protects them from overcharging. These same back-end components could also store power from a solar cell array, a micro-hydro turbine, or any other off-grid, environmental power source.

BATTERY

Tim Lillis

Mini Jacob's Ladder

Written by Heinz Behling, Translation by Niq Oltman

TIME REQUIRED:
1–2 Hours (Plus 5 Hours if Printing)

DIFFICULTY:
Easy

COST:
$15–$20

MATERIALS
» **High voltage generator, input 3.6V–6V, output 20kV** aka igniter or step-up or boost power module, AliExpress #32845852439
» **3D printed enclosure (optional)** Download the free 3D files at thingiverse.com/thing:3182601 or github.com/Make-Magazin/Jakobsleiterchen (see text), or fabricate your own enclosure out of nonconductive materials.
» **Power supply** You've got two options:
  ▪ **AC wall adapter, 5V, minimum 4A, 5.5mm/ 2.1mm DC plug** such as Adafruit #1466
  ▪ **5.5mm/2.1mm DC barrel jack** Adafruit #373
  — OR —
  ▪ **Batteries, 3.2V, lithium iron phosphate (LiFePO$_4$ or LFP), size AA (2)** such as AliExpress #32885193856. Don't use regular lithium ion (Li-ion) batteries; their output voltage is too high.
  ▪ **Battery holder, 2xAA** AliExpress #32860258429
  ▪ **Charger** for LiFePO$_4$ batteries, AliExpress #32762060735
» **Switch, mini pushbutton, momentary, rated 0.5A 24VAC or higher, ¼" mount** such as RadioShack #2751556
» **Solid wire, 0.8mm–1mm diameter** for the ladder electrodes. Previous builds have used silver-coated copper wire; I used 1mm zinc-coated steel wire.
» **Hookup wire** for the circuit. Use copper wire at least 0.5mm² if you're using the LiFePO$_4$ cells.
» **Neodymium magnet, 8mm diameter, 5mm high**
» **Terminal strips, screw type**
» **Empty jam jar, 60mm opening, about 9cm high**
» **Glue, heat resistant** such as epoxy

TOOLS
» **Soldering iron and solder**
» **Wire cutters and strippers**
» **Screwdriver**
» **3D printer (optional)**

HEINZ BEHLING is an editor for *Make:* Germany in Hannover who's into 3D printing and laser cutting. He started out long ago on the Commodore VC20 and C64; today he's 60 but still not an adult.

This pocket-sized ladder is easy to make for just $20. You'll be rewarded with a real high-voltage eye-catcher.

In times past, Jacob's ladders were often featured in horror flicks for their decorative effect. As the bright purple spark climbs between the electrodes, growing longer as it rises, it crackles with a sound that says "mad scientist"! As a spark gap, the Jacob's ladder also makes a very good high voltage surge arrester, and they're still used for this purpose today, such as on overhead wires for trains.

If you're looking for a home-friendly version, you'll find one on Thingiverse. Created by Matthias Balwierz, aka bitluni, it's all packaged inside a 3D printed enclosure; you can see it in action in Matthias' video at youtu.be/0Dp51Z-iPF4. More details on this electrifying project follow below. But first, let's have a look at how the ladder works.

DANGER: HIGH VOLTAGE! This little ladder generates very high voltage, up to 20kV. It is unsafe for children or users of pacemakers or similar implants.

Heinz Behling

How the Jacob's Ladder Works
The graph shows the voltage level at the electrodes during a cycle.

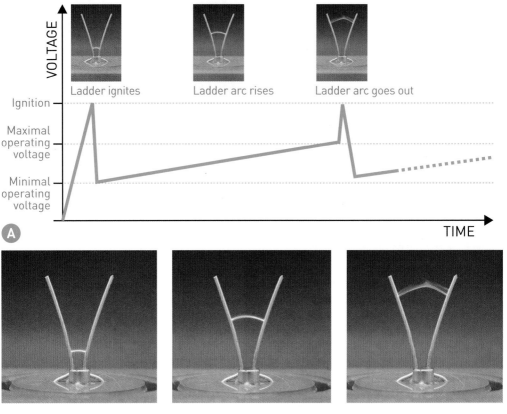

Ladder ignites   Ladder arc rises   Ladder arc goes out

Ignition

Maximal operating voltage

Minimal operating voltage

VOLTAGE

TIME

Ⓐ

Heinz Behling

LADDER LOGIC

The functional principle is pretty simple (Figure Ⓐ). A voltage is applied to the ladder electrodes from a transformer — say, from a neon sign, microwave oven, or model train, or, in our case, the output from our little high voltage generator. If that voltage is high enough, an arc forms where the gap between the electrodes is shortest. (For an arc to "make the jump" across the gap, you need about 1,000 volts per millimeter.) This arc is simply air that has been ionized by the voltage, making it electrically conductive. The electrical energy flowing across the arc is partly converted to light, heat, and magnetic fields. This causes the voltage across the electrodes to drop significantly, as the resistance of the arc provides a load on the high voltage source.

The arc's heat — and, to some extent, its magnetic field as well — cause it to travel upward. At this point, the voltage is too low to

ignite another arc. As the existing arc travels up, the widening gap between the terminals forces it to stretch. Recall that wider spark gaps require higher voltages for arcing: the arc keeps traveling until the voltage can no longer sustain it, at which point it breaks down. With the arc gone, there's no load on the voltage source anymore, and the voltage will again begin to rise. Once it's high enough for a new arc to form, the cycle repeats.

Enough theory; let's move on to practice. For starters, a word of warning: the Jacob's ladder requires high voltage. The voltage boosting module we use can supply up to 20 kilovolts! Use great caution when working on this project — avoid any contact with the voltage. Don't build this project if you wear implants such as pacemakers, insulin pumps, or similar. This build is not safe for children, either.

The parts are cheap. Including the power supply (power brick or LiFePo batteries), you'll

need maybe $20 worth of materials. The build can be completed in about an hour. If you're going to 3D print your own case, this will take longer (around 6 hours), but it's not necessary. In principle, you can use any enclosure made from non-conducting materials. You may need to get creative figuring out how to attach the terminals and the glass hood.

1. MAKE YOUR ENCLOSURE

3D files for printing the enclosure can be downloaded at thingiverse.com/thing:3182601 for the battery version, or github.com/Make-Magazin/Jakobsleiterchen for our power brick version. The top of the enclosure should face the print bed. Print with 20 percent infill and supports added.

The enclosure contains a ring-shaped socket that fits the opening of a small empty jam jar turned upside down (Figure B). This jar will serve

as a protective hood, keeping onlookers safe from the high voltage. It also provides a draft shield for the ladder: because the rising motion of the arc is caused primarily by its own heat, it's sensitive to air currents.

2. WIRE THE CIRCUIT

For the ladder electrodes, the only suitable material is solid wire. The original build by Matthias uses 0.8mm silver-coated copper wire. I've used 1mm zinc-coated steel wire, which is harder to bend to shape, but this also makes it more rugged against accidental bends. Start with two pieces about 8cm long; you'll trim them later. We recommend attaching these electrodes to your circuit using ordinary screw terminal strips; heavy wire is hard to solder securely, and zinc-coated wire is impossible.

Secure the high voltage boost module and the terminal strips inside the enclosure using some

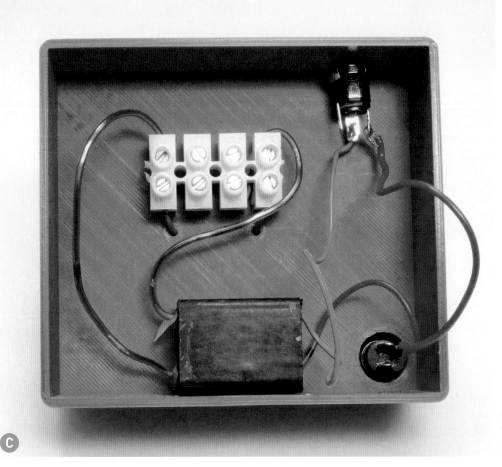

Heinz Behling

glue, making sure there's still enough room for the battery holder if you're using batteries.

Wiring it up is easy (Figure **C**). The red wire from the high-voltage boost module is positive (+), the green one is negative (–). Connect the red wire via the pushbutton to the positive (+) output of the power supply. Connect the green wire directly to the negative (–) output of the power supply. The two wires on the opposite end of the boost module conduct the high voltage; connect these to your ladder electrodes via the terminal strips.

For the power supply, you can use two AA-size LiFePo cells (3.2V each) in a suitable holder, plus a matching charger. (Note that regular Li-Ion batteries can't be used here, as their output voltage is too high.) The cheapest solution is to use a power brick (5 volts, minimum 4 amps, Figure **D**). If you're going for the latter, use our modified 3D printing file for the enclosure that has an opening to fit a power jack for the brick.

> **NOTE:** In Matthias' build, there's also an option to power the device using four plain *alkaline* batteries, but we couldn't get this option to work; the current delivered by the batteries was too low.

3. BEND AND CUT THE LADDER

Once you've completed the assembly, it's time to bend the ladder electrodes into the proper shape. At the base, the gap should be about 4mm to 7mm wide. The electrodes should reach up to a height of about 4cm. At the top, make the gap about 2.5cm wide.

4. TEST AND ADJUST

You can now connect the power supply and start the ladder by pushing the button. Observe the spark gap. You probably won't get the desired effect right away. In that case, you need to adjust the electrodes by bending them.

> **⚠ DANGER: HIGH VOLTAGE!** Always disconnect the device from the power supply before touching the electrodes! (Should you accidentally push the start button, your body may act as the resistive load instead!) When using a power brick, it's not safe to just unplug it from the wall socket. The capacitors inside the power brick retain enough energy for at least one more ignition of the arc! Instead, always unplug the output side of the power brick from the connector jack on the Jacob's ladder.

Here's how to adjust the electrodes depending on what you're seeing:
» If there's no spark at all, the bottom gap is too small.
» If the ladder ignites but the arc fails to travel upward, instead dancing around the base where the gap is smallest, make sure the electrodes are the same shape and height.
» If the arc travels up but doesn't break down (extinguish), the top gap is too small.

Don't despair if your Jacob's ladder turns out to be stubborn. We have ways to deal with this! The effect works best if the electrode wires

are completely smooth. Try cleaning them, or scrubbing them with very fine steel wool. Rubbing them with the tip of a soft pencil lead (graphite) can also improve the effect.

5. MAGNETIC MOTIVATOR

Under some conditions of temperature and humidity, the arc will simply refuse to move, as the heat it produces is insufficient. So we'll take advantage of the fact that it generates its own magnetic field: we can use an external magnetic field to make it budge! Placing a small permanent magnet underneath the electrodes (south pole facing up) will literally push the arc away, setting it in motion. Since this works so well, we also provide a modified 3D print file that's got a recess for fitting a small neodymium magnet (8mm diameter, 5mm high). Glue the magnet in place (Figure **E**) to keep it from sticking to steel wires.

SWEET BABY JACOB!

Your mini Jacob's ladder is now ready for use. Before you go wild, note that it's not made for permanent operation: the high voltage generator will begin to overheat after about 1 minute of continuous use, which will shorten its lifespan considerably.

The arc also produces some ozone (O_3) and nitrogen dioxide (NO_2), which can be very hazardous, particularly if allowed to accumulate. We strongly recommend that you operate the ladder only with the glass hood mounted, and ventilate the room after use. Have fun! ❷

Share your build at makeprojects. com, and watch Matthias Balwierz' original build video at youtu. be/0Dp51Z-iPF4.

Sequino!

How I made a unique "rewrite" clock with Arduino, CNC, and sequins

Written and photographed by Ekaggrat Singh Kalsi

EKAGGRAT SINGH KALSI is an architect in Beijing who graduated from CEPT in Ahmedabad, India. In his free time he tinkers with electronics and mechanics and takes full advantage of an electronics market 5 minutes' walk from his home, and of course Taobao online. ekaggrat.com

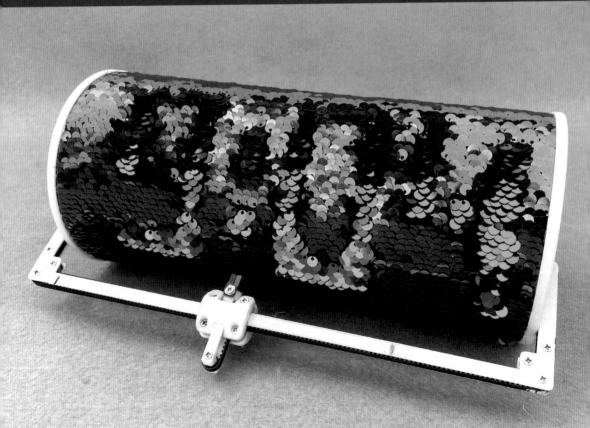

TIME REQUIRED: Several Weekends
DIFFICULTY: Advanced
COST: $40–$60

MATERIALS
» **Sequin cloth**
» **Foam, 3mm, open cell**
» **3D printed parts** Download the 3D files for printing at thingiverse.com/thing:4312491.
» **Arduino Nano R3 microcontroller**
» **Nano CNC shield** such as Keyes CNC, Amazon #B07BGVST9D
» **Stepper motor drivers, StepStick A4982 (3)** such as Amazon #B01KULNLDG
» **Hall sensor**
» **IR reflective sensors (2)**
» **Stepper motors, 24BYJ48, 1:64 (4)**
» **Timing belt, 2GT, 3mm wide** I split a standard 6mm belt from a 3D printer in half.
» **Bearings, 602 (12)**
» **Assorted screws and hardware**

TOOLS
» **3D printer**
» **Soldering iron**
» **Screwdrivers**

This project is a continuation of my quest to build a robotic clock that can write and rewrite the time continuously, day in and day out. My first attempt, Doodle Clock, was a failure due to the marker drying up. Doodle Clock #2 failed because the display — a kids' magnetic drawing board — soon got scratched.

This new clock, Sequino, was inspired by my daughter's T-shirt with the pattern-changing sequin cloth glued to it. After getting some stock sequin fabric from a vendor, I figured out the size of the clock based on the minimum resolution of the cloth: it has 5mm circular sequins stitched 3mm apart. Another limit of the clock was my 3D printer bed size of 245mm×170mm.

THE KINEMATICS
The kinematics of the clock consists of a mix of H-bot and Cartesian systems (Figure A). The y-axis is a hubless axis consisting of two rings on the extreme ends of the clock (Figure B) driven by two 24BYJ48 motors. The rings are basically gears constrained in outer rings, similar to a hubless wheel.

The x-axis consists of a strut holding the two

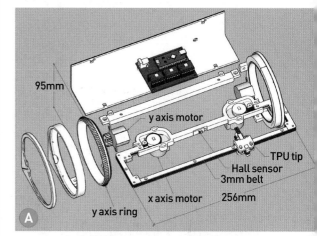

95mm · y axis motor · TPU tip · Hall sensor · 3mm belt · x axis motor · 256mm · y axis ring

A

B

rings together and driving a belt routed on a H-bot based path (Figure C on the following page). The belt is driven by two motors; when these move in the same direction, the pen carriage moves left or right on the x-axis, and when they move in opposite directions the belt tension moves the "pen" up or down (z-axis) to engage with the sequins.

The y-axis homes to a Hall sensor on the backside and the x-axis homes to an optical sensor on the right side with the help of a white mark on the rubber belt. The pen homing is tricky. After a lot of trial and error I found a way out, by first homing the x-axis, then backing up a bit and then homing the pen at a known belt mark to a sensor on the left side of the clock (shown in Figure E on the following page), and then

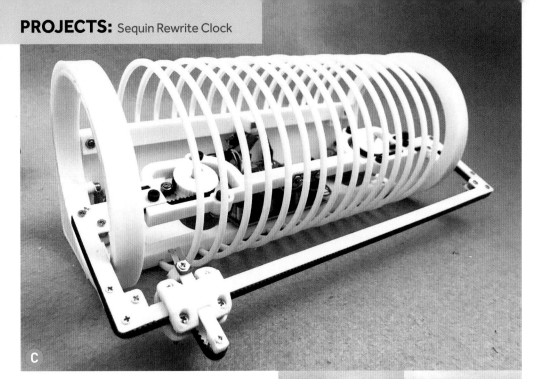

finally moving a fixed distance to zero it. This isn't perfect, as it sometimes overshoots the mark, but an error of 1mm on the pen is OK in this case.

FIGURING OUT THE CLOTH BACKING

The sequin cloth consists of shiny sequin discs that are stitched to a fabric backing. The sequins, colored pink on one side and blue on the other, flip easily with a finger when they're mounted on a T-shirt or a bag with a soft backing. But the moment I mounted it on a cylinder it stopped flipping as the discs got over-constrained by the hard backing. First I tried a ribbed structure as a backing with the fabric stretched on it. This worked but was very complicated to construct.

Finally I figured it out that a simpler way was to just add a 3mm sponge foam layer as a backing to the cloth to give the sequins the freedom to flip (Figure D).

THE TIP OF THE PEN

The tip of the pen was very tricky to figure out. Getting a material to act like the tip of a human finger is not simple. The disks are very slippery and need just the right friction to flip them. The final solution was a tip made of TPU plastic (thermoplastic polyurethane) with a split hook (Figure E).

Sponge backing   Ribbed backing

Optical sensor and belt mark   TPU pen tip

WHAT'S NEXT

Sequino is still a prototype, although I've tried to use maximum off-the-shelf components. It can be built by an advanced user — the 3D printing is a bit tricky and the optical sensors need a little bit of work. I intend to make it a kit sometime soon, and to share an Instructable, but I've been a bit stuck as I am in Beijing and things have been held up due to the outbreak! ✪

Watch the Sequino clock in action at youtu.be/6jpsilsMG8U.

You can find them all at ekaggrat.com and get build details for most of them at hackaday.io/ekaggrat.

1

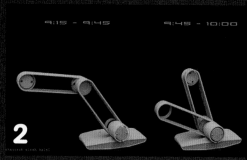

9:15 – 9:45　　　　9:45 – 10:00

2

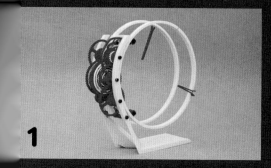

9:08

3

4

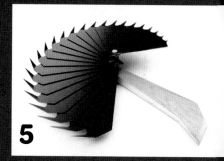

5

1. TORLO

The Torlo was born out of the idea of using a simple oscillating motor as a power source. The voice coil of a scrap laptop hard drive fit the bill nicely; it's pulsed by an ATtiny2313 every 2 seconds to drive the balance wheel, which pushes a cam and a ratchet to turn the clock 2 seconds further. The rest of the clock is a simple drive train driving the minute and hour rings, which display the time.

2. EDGYTOKEI

The Edgytokei ("edge clock") is inspired by Japanese nunchucks — just a pair of arms displaying the time by balancing themselves on edge. Both arms are of equal length, as their roles change with different hours of the day. The fulcrum of the clock flips from the center to the left or right every quarter hour, so that the clock can always stand on edge. Both arms have LEDs; whichever one represents the hours lights up.

3. DOODLE CLOCK #2

I built my first Doodle Clock with marker pens as a joke, but watching it work was so mesmerizing that I wanted it to be a practical desk clock. The problem was the markers dried up after 30

minutes. So the solution I found was magnetic writing boards made for children. I used 2mm cylindrical magnets inside a solenoid to write and erase the text, and small geared steppers to make the clock silent and smooth. The clock is run by an ATmega644p with Arduino bootloader, the motors are run by the standard StepStick drivers, and the coils are run by a 1293DD dual H-bridge. The kinematics for the arm were solved by a user on the RepRap forum!

4. HOLOCLOCK

My first 3D printed clock, based on a single geared stepper motor I found in surplus, 10 for a dollar. It's driven by an ATtiny2313 and ULN2803 Darlington circuit, with the code written in Arduino IDE. The ATtiny pulses the motor every minute to move a gear train, which in turn moves the minute ring and hour ring. I still sell it as a kit at tindie.com/products/ekaggrat/3d-printed-holo-clock.

5. SPIRE

The spiral form of this clock unfolds and folds in the rhythm of a Japanese fan. This project, co-designed with Darshan Soni, won a Red Dot Design Award.

Comprehending Capacitors

A capacitor stores electricity, as you will see in this experiment

Written and illustrated by Charles Platt

Some capacitors have colored cans. Others don't. The color is not important.

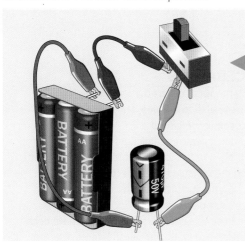

I wrote my book *Easy Electronics* to help beginners get acquainted with electronics more simply, quickly, and affordably than ever before. A dozen hands-on experiments show you the basics, and each one takes half an hour or less. You won't need any tools *at all*.

You will need: AA batteries (3) with holder, alligator jumper wires (3), 470µF capacitor, 100µF capacitor, 1kΩ resistor, through-hole 5mm LED, SPDT slide switch.

EXPERIMENT:

The type of capacitor shown here is called **electrolytic**. Its storage capacity, known as **capacitance**, is 470µF—but I'll explain that in a moment. 50V is its maximum voltage, but for this experiment, a rating of 10V or higher is okay.

The short lead is the negative side, also identified with minus signs. That's because this capacitor has **polarity** — never connect an electrolytic capacitor to a power supply the wrong way around.

You can build this circuit in two steps. This part just charges the capacitor with electricity when the slide switch moves to upper-left.

Some of the voltage from the battery transfers to the capacitor, although you can't see any sign of it yet.

Add the 1K resistor and the LED, with the negative side of the LED sharing the negative leg of the capacitor.

Now move the switch to the lower-right. The capacitor discharges itself through the LED.

Move the switch to the upper-left and wait 5 seconds for the capacitor to recharge. Now you can discharge it again and light the LED again!

If this diagram looks complicated to you, try sketching a copy of it, replacing the alligator wires with simple lines to connect the components.

CHARLES PLATT is the author of *Make: Electronics*, an introductory guide for all ages, its sequel *Make: More Electronics*, and the 3-volume *Encyclopedia of Electronic Components*. All these and his workshop guide, *Make: Tools*, are available at makershed.com/platt.

HOW DOES IT WORK?

Inside the capacitor you used are two pieces of metal film known as **plates.** They are separated by paste called an **electrolyte,** which is why this capacitor is called electrolytic.

When electrons flow into one plate, they try to create an equal, opposite charge on the other. You can think of the plates as having positive and negative charges that attract each other.

TIMING

The 1K resistor was needed because you charged the capacitor with 4.5V from the battery pack, and the LED can only handle about 1.8V. The resistor prevents the LED from being damaged.

The resistor also controls how fast the capacitor discharges. Substitute a 10K resistor (brown, black, orange) and the LED is dimmer than before and takes much longer to fade out.

Here's another thing to try. Go back to using the 1K resistor. Remove the 470µF capacitor and substitute a 100µF capacitor. Push the switch to and fro, and now the LED lights up very briefly.

Electricity moves fast, but a capacitor and a resistor can make things happen slowly.

CERAMICS

Capacitors such as the one shown above are less than ½" wide. They are dipped in a **ceramic** compound.

Most ceramic capacitors do not have polarity.

Many ceramic capacitors have a code printed on them instead of their actual capacitance.

Some ceramic capacitors are shaped like circular discs.

In simple circuits of the type you have been building, usually you can substitute a ceramic instead of an electrolytic if you wish. Note that for values around 10µF and above, ceramics may be more expensive.

UNITS

Capacitance is measured in **farads**, abbreviated with letter F. But a 1F capacitor is very large. In hobby electronics we mostly use capacitors rated in microfarads, abbreviated µF. The µ symbol is the Greek letter mu, but often µF is printed as uF.

There are 1,000,000 microfarads in 1 farad, 1,000 nanofarads (nF) in 1 microfarad, and 1,000 picofarads (pF) in 1 nanofarad.

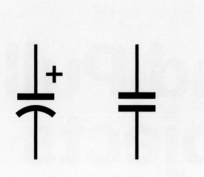

SCHEMATICS

There are two symbols for capacitors.

A **polarized** capacitor, such as an electrolytic, is on the left.

A **nonpolarized** capacitor, such as a ceramic, is on the right.

Some people use the symbol on the right everywhere in a schematic, and let you decide if you want to use an electrolytic capacitor, and if so, which way around it should be.

YOU MIGHT THINK:

A capacitor may seem similar to a battery. After all, they both store electricity.

A battery, however, uses chemical reactions, and even a rechargeable battery wears out after a limited number of charging and discharging cycles.

A capacitor does not use chemical reactions, and can still work as well after several years.

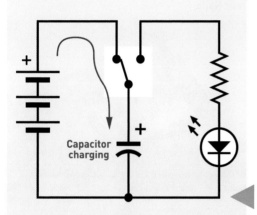

In this schematic showing the circuit that you just built, the double-throw switch has completed a circuit with the battery, so that the battery charges the capacitor.

In this schematic, the double-throw switch is in its other position, completing a circuit from one plate of the capacitor, through the resistor and the LED, back to the other plate. ✏

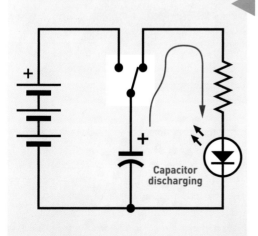

This article is adapted from the book *Make: Easy Electronics*, available at makershed.com/platt and other booksellers. A complete kit of all components used in the book is available at protechtrader.com/easyelectronics.

Pick and Pull PC Projects

Here are three fun projects I've built with my salvaged computer parts

Written by Samer Najia

When I first started my IT consultancy years ago, I set up a series of computers to function as servers, and a variety of dedicated workstations to test software. For the most part, each of these machines was hand built and configured to suit. Then one day I moved them all to a server facility and virtualized them, and suddenly I had a whole bunch of high-power computers eating up a ton of room in my garage! It was time to either tear them apart or sell them — or better yet, repurpose the best of them.

I'M BURIED IN COMPUTERS

I knew well enough how to take computers apart, but multiple trips to Take-Apart events had also taught me the innards of laptops and printers and peripherals, and I was harvesting LCD panels, drives, and power supplies. It wasn't long before my friends were handing me their old equipment to harvest parts from. I had bins full of all sorts of goodies for building network storage devices, media servers, Linux boxes, and so on. And I wanted to do something interesting with it all.

I started working on a couple of projects that were near and dear to my heart, and both were going to need computers. Powerful computers. Of course, computers don't need to live in a PC case to be useful, so I've let myself be creative with their integrations. Here then, are a few of my really cool uses for old computers.

PROJECT #1: 3D Printer with Embedded Computer

I always hated the idea of tying up my laptops to run my 3D printers. There are printers that are effectively standalone machines that can do their own slicing and processing and then of course controlling. And sure, a single-board computer like a Raspberry Pi can do some of that, but I wanted my project to really be standalone and do everything. I decided I needed to embed a full-blown computer into my printer — and why not? I had more than enough parts. To me, the whole idea was all pros and no cons:

» The 3D printer no longer needs a separate computer and can do its own slicing using any software that runs on Windows (this is not really as easy to do with Macs)

MATERIALS

Scavenged computers and peripherals
These yield lots of useful parts:
» **Motherboards**
» **Memory** on the motherboard
» **Hard drives**
» **Power supplies**
» **Fans**
» **Power and reset switches** You can buy standalone switches that plug right in to the motherboard from various vendors on Amazon and eBay if you can't remove them from the computer case
» **LCD displays**
» **Screws and brass standoffs**
» **Keyboard and mouse**
» **Assorted OEM driver boards, USB dongles, and peripherals** depending on your project
» **Aluminum extrusion and brackets, 3D printed parts, plywood, acrylic sheet, veneers, etc.** for making custom enclosures

TOOLS
» **Jigsaw**
» **Screwdrivers**
» **3D printer (optional)**
» **Allen key sets**
» **Wrenches, socket and standard**
» **Ruler and measuring tape**

Adobe STock - damrong, Samer Najia

SAMER NAJIA holds a degree in mechanical engineering from Duke University but making things is his true passion. He spends countless hours building progressively larger and more complex projects. He's co-author of the new book *Mechanical Engineering for Makers: A Hands-on Guide to Designing and Making Physical Things*, available at makershed.com/books.

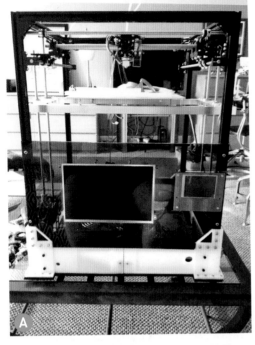

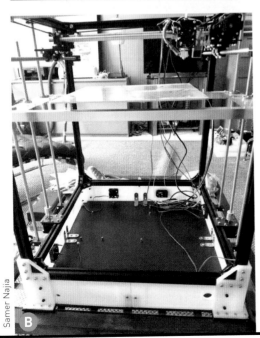

A

B

Samer Najia

» Remote control of the printer is simply a Remote Desktop connection
» It can act as a server and be accessed remotely using a web browser (with some configuration)
» I can run just about any file or software I want and worry less about processing power
» I can upgrade software and hardware anytime, add peripherals and capability, even replace the motherboard and reconfigure the operating system with little to no loss of the installed on-board software.

The printer I chose for this project is an FLSun Cube — an H-bot style printer with a larger print bed. It's fabricated from 80/20 type extrusions and, like most PCs, is essentially a box. With a few modifications to extend the chassis, it can host a motherboard, ATX power supply, hard drive, fans, and wiring. Here were some criteria for the computer parts I chose:

» The motherboard had to have an onboard video display card and preferably HDMI, because I didn't want to consume space with a vertically standing video board in a PCI slot.
» The computer had to have at least 5 onboard USB ports: for the 3D printer itself, a Wi-Fi dongle, a wireless keyboard dongle (if your board already has Wi-Fi and Bluetooth radios on board, even better), an external storage device (just in case), and a combination wired keyboard and mouse. I had several laptops that fit the bill, but preferred to empty my parts bin.
» I wanted to use a smaller harvested LCD with a commercial OEM controller board (which means eBay or Amazon it seems these days). I used an 11" LCD pulled out of an old notebook computer. I could just as easily have purchased a standalone HDMI-based LCD (even a touchscreen), but again, parts bin...
» The hard drive needed enough storage to install all sorts of software and of course to store any printer-specific G-code or 3D models. I used one from my old servers.

The outcome: a completely standalone computer/printer hybrid (Figure Ⓐ). Notice the white 3D printed parts that raise the printer up enough to build a box underneath; these are from various designs I found on Thingiverse and modified

slightly to suit my purposes.

Figure **B** shows the printer's interior with side panels and fans installed. The floor of the box is plywood, cut to suit and covered with a piece of lamination meant to look like carbon fiber. You can see the motherboard standoffs being installed. Notice the rear panel containing the power socket.

In Figure **C**, the motherboard is installed, oriented correctly and ready for the rest of the components. It was important to test-fit first and consider how the cables lay out and how easy it is to install wiring and plug or unplug things.

And here's everything installed in the printer base (Figure **D**). Notice the two power supplies (for computer and printer) and the hard drive in the back, as well as how the SATA and USB cabling are routed. There are also two power plugs in the back of the printer.

PROJECT #2: Custom Flight Simulator

I wanted to upgrade my home-built flight simulator to have more power and feel more like a real airplane. That meant building an panel, seat, and controls that are close enough to the real thing to feel like it. I had a bunch of instruments from the previous iteration, a few old LCDs, and a bunch of 80/20 extrusions. And of course, I had some of those computers.

I wanted my simulator to be able to do most of what the expensive desktop simulators do, and to be upgraded at will. The computer had to support additional instruments and aircraft models, which might mean more monitors and more USB ports. I also had plans to add more computers to the system to run more instruments and controls. Finally, the simulator had to be mobile enough to roll around and fit through a standard doorway.

Once a frame was assembled, the next step was to come up with a panel layout. I arranged the purpose-built instruments in a logical sequence (radios and GPS to the right, main flight instruments to the left) with a harvested LCD as the main instrument display so that I can show analog gauges or a moving map. Controls were arranged for the pilot of a fixed-wing aircraft (left seat in the cockpit, with throttle controls on the right). I used a boat seat I found on Amazon, and because I wanted it to be adjustable I made a rail system for it using V-slot wheels and parts from an old CNC machine.

One critical decision: where to park the computer and its parts? I found myself constantly opening and closing the case as I changed video boards, added USB capabilities, changed the power supply, fried the hard drive and had to replace it. I finally chose to integrate the computer into the simulator itself so that when I upgrade it, I won't have to revisit all the wiring and space allocations. I mounted the motherboard, hard drive, and power supply directly under the seat so I can change out parts anytime, while still encasing them. The box under the seat can be covered by a sheet of acrylic while leaving the bottom open to allow airflow.

Here's the simulator at its home at my flight school (Figure **E**, next page). You can just about make out the motherboard toward the bottom of the image, and the power supply is to the right.

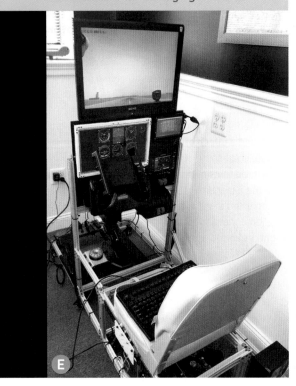

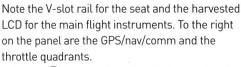

E

F

Samer Najia

Note the V-slot rail for the seat and the harvested LCD for the main flight instruments. To the right on the panel are the GPS/nav/comm and the throttle quadrants.

Figure F is a view from the right side, showing a better rendering of the GPS units and radios, the rudder pedals, and the structure of the frame around the panel. The panel itself is ¼" MDF with cutouts made with a jigsaw.

Figure G shows the mounted motherboard. In the foreground is an SSD drive mounted on a PCI board and behind it is a video board supporting as many as 6 monitors. All the peripherals were purchased on Amazon and were specific to my needs. If you choose to build one of these, your needs may vary and the combinations are mind-boggling. Do consider keeping as much on the motherboard as possible and minimizing the number of add-on cards.

Figure H is another view of the motherboard. The connections were pretty simple and the number of add-on cards was kept to a minimum. I purchased an "open case" kit to help hold the cards in place and then mounted the whole thing on the 80/20 extrusions.

The kit allows for a large fan to push air over

the board (Figure I), but I could have mounted the same fan to the side of the extrusion.

Figure J is the next version of the simulator panel — now that the first one is at the flight school, I need one for home. Notice the Arduino-based instruments toward the bottom and the harvested LCD at top left. Those are all available from a variety of vendors. These days more and more GPS/nav/comms with onboard transponders are available as touchscreen devices so if you set up a TFT screen as the instrument, the rest is all done in software.

PROJECT #3: DIY Garage Monitors

As mentioned before, the first thing I pull out of a laptop is the LCD, and if I get the speakers out as well, so much the better. Sure, monitors are cheap and very common (and heavy) but I try to eliminate e-waste by harvesting all I can.

Now, not all LCDs are created equal and sometimes I can't find a driver board, but most of the time I can, either on eBay or on Amazon. You don't need much really. Most of the work is developing and assembling an enclosure. I've made several, some with 10mm extrusions, some with 3D-printed panels, and others with pieces

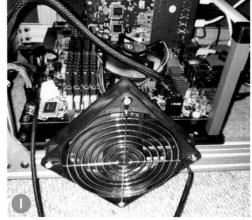

of cut acrylic. Be as creative as you like. Here are two examples to get you started:

» Figure **K** shows a simple monitor with 10mm extrusions (3D printed versions, but it's just as easy to cut metal ones). The back is just plywood (Figure **L**).

» The monitor in Figure **M** has a mixture of aluminum 10mm extrusion and 3D printed corners. The panels can be printed or cut from sheet styrene. The back of this monitor is a clear piece of acrylic. This unit has its own battery pack, charging unit, speakers, and VESA mount. There's a Raspberry Pi in there too (Figure **N**).

I hope you find these ideas interesting. Let your own imagination run and you'll come up with all sorts of ideas. I used to think that once a computer was no longer useful it had to be sold or trashed. Now I look at each one as a source of parts for my bin and potentially a component for one of my many upcoming projects. ◗

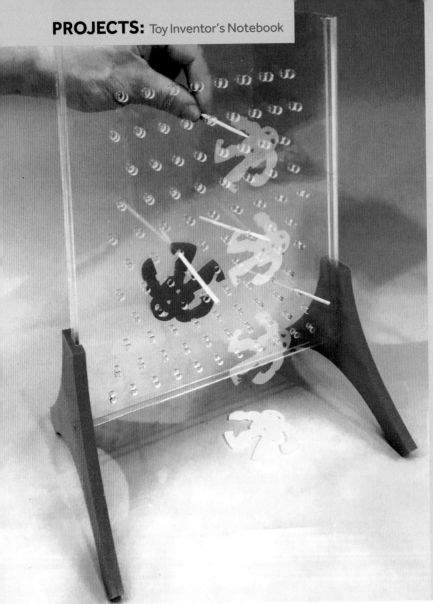

TIME REQUIRED:
1–2 Hours

DIFFICULTY:
Easy

COST:
$20–$40

MATERIALS
» **Acrylic sheet,** ⅛" thick
» **Acrylic cement** such as methylene chloride (MC), with applicator
» **Wooden skewers** Any ol' kitchen skewers will do — the ones I used are 6"×³⁄₁₆" or so. No dimension is critical. You could use small wood dowel, or pieces of stiff wire, or spaghetti ...

TOOLS
» **Laser cutter (optional)** I used a Glowforge, but you can also cut and drill the parts by hand.
—OR—
» **Fine-toothed saw**
» **Drill**
» **Measuring tape or ruler**
» **Clamps or tape**

Don't Drop Your Bot!
High-tech fabrication meets low-tech play action Written and photographed by Bob Knetzger

BOB KNETZGER is a designer/inventor/musician whose award-winning toys have been featured on *The Tonight Show*, *Nightline*, and *Good Morning America*. He is the author of *Make: Fun!*, available at makershed.com and fine bookstores.

Ⓐ

Here's a classic skill-and-action game updated for you to make and play. You and your opponent each start your robot at the top of the playfield. Use the skewers to hang your bot in place. Ready, set, go! Race to remove and replace the skewers one by one to lower your bot down, down, down … but if your bot slips off the skewers and falls to the bottom — oops! — you must start over at the top. The first player to get any part of their bot hanging down below the field wins (Figure A).

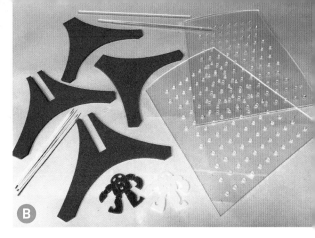

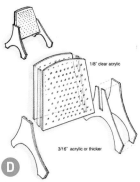

1/8" clear acrylic

3/16" acrylic or thicker

BUILD THE GAME

For this game, make 2 of each part (Figure B). They're symmetrical, so you can stack 2 pieces of acrylic together and then just measure, cut, and drill once to make 2 identical parts. Of course making this is super easy with a laser cutter! Find the bot image, vector art, and .svg files online at makezine.com/go/toy-inventor-73.

To make the bot game pieces, I used one of the unique features of the Glowforge: Its built-in lid camera scans your hand-drawn artwork, then it cuts and engraves the part, no .svg file needed — great for whimsical shapes like this (Figure C).

Rather than fitting together delicate slots in brittle acrylic, the parts are simply solvent-bonded with acrylic cement (Figure D). This gives a much stronger design that's easy to assemble. First, bond the side spacers to the playfield panels (Figure E). Bond the leg panels together (Figure F), and then lastly bond the two legs to the completed playfield (Figure G).

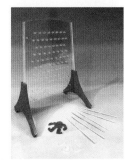

PLAY IT

» Start with three skewers for a challenging game: two skewers to start the bot at the top (Figure H) plus that third skewer to help catch, hold, or trap your bot as you try to move it down.

» Add more skewers to make it easier: put them in the bottom row of holes to catch the bot, then strategically place/remove skewers to swivel the bot to cross the finish line without falling.

» Choose your strategy: make cautious, incremental moves, or be bold (Figure I). A single, daring drop to a well-placed skewer below could win the game in one move!

» Give younger players a head start to handicap your races.

» Whatever you do, don't drop your bot! ⊘

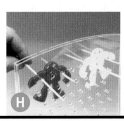

1+2+3 Mini Ground Effect Vehicles

Written and photographed by Doug Stith

Experiment with the world's lowest-flying aircraft

TIME REQUIRED:
30–60 Minutes

DIFFICULTY:
Easy

COST:
$1–$2

MATERIALS
» **Paper cardstock** such as index cards
» **Mini binder clips**
» **Small paper clips**
» **Tape**

TOOLS
» Ruler
» Pencil
» Scissors

DOUG STITH is a 33-year veteran middle school science teacher who loves creating things for his students in his workshop. He wrote "Rumblebots Raceway" in *Make:* Volume 52 (makezine.com/projects/rumble-bots).

Ground-effect vehicles (GEVs), also known as wing-in-ground-effect vehicles (WIGs) or *ekranoplans*, are craft designed to fly low over flat surfaces. Air is trapped between the wing and the ground, and the GEV rides on a cushion of air; that's the *ground effect*. People sometimes confuse GEVs and hovercrafts. Hovercrafts propel air downward to achieve lift; GEVs don't. Hovercrafts need no forward velocity to attain lift; GEVs do.

The ground effect is most pronounced very close to the ground. For safety reasons, the ground should be level, which is why flight over water is the most common use of GEVs. One advantage of GEVs over conventional planes is the lack of wingtip vortices that produce drag.

Pilots discovered the ground effect in the 1920s, when they found that their aircraft became more efficient as they neared the runway. It was another 40 years before working GEVs were built. Rostislav Alexeyev in the Soviet Union and German Alexander Lippisch, working in the United States, were leaders in GEV design.

Last fall I stumbled onto John Ryland's YouTube videos of GEVs (youtu.be/TZ4U89POYNc). Cool! I had never heard of them. I searched and found more designs (youtu.be/JcUMO6xken8 and instructables.com/id/Paper-ground-effect-vehicle). As a science teacher, I was more interested in these simple homemade versions than actual working crafts.

1. Make the GEV body

I tried simplifying the designs of others. I had good luck with a body made out of one piece of cardstock and a rear stabilizer made out of a second. Although I varied the dimensions, this is what I began with (Figure A).

Fold along the dotted lines (Figure B) at a 90° angle to produce the final body shape (Figure C).

2. Add the stabilizer

Next, a rear stabilizer is needed, and not just for looks. Cut it out (Figure D), fold on the dotted lines (Figure E), and tape (Figure F).

Now attach the stabilizer to the body (Figure G).

3. Add clips for weight

Place your vehicle on smooth surface like a large table or floor. Place your hand at the back of the vehicle and give it a quick push. You'll find that air will flow under the body and push the front up. Weight is needed in front. Mini binder clips work well (Figure H).

Flying Your GEV

If the front still lifts too much, add more weight. And if the vehicle won't lift off the ground, there's too much weight. In addition to binder clips, try small paper clips; sliding them forward and backward slightly produces noticeable effects. When you've got it right, your GEV will glide surprisingly far on its cushion of air! ⊘

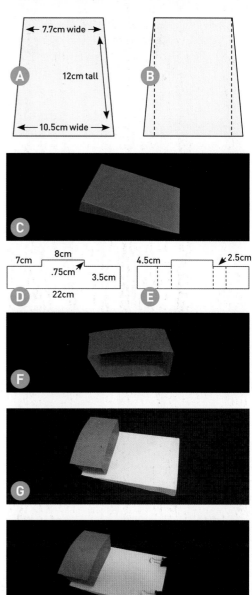

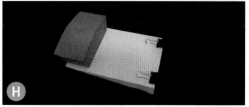

See more mini GEV experiments to try at youtu.be/Um80kldsOYO

Green Grain
GLAZES

Study up on quality woodworking finishes that won't destroy the world

Low-VOC finishes I use in my shop.

TotalBoat tries to save on packaging and waste with its bagged instead of canned Halcyon finishes.

Rubio Monocoat: This little soup-size can costs as much as a gallon of other finishes, but a little goes an incredibly long way.

TIM SWAY is a Connecticut-based artist and maker who specializes in reclaimed, upcycled, and eco-friendly woodworking. His mission statement is to "Make Worthless Things Priceless" and he's currently focused on making affordable and eco-ethical guitars. You can learn more about him at youtube.com/TimSway, NewPerspectivesMusic.com, and timsway.net.

You build tables from reclaimed beams, using solar-charged tools by daylight. Heck, you even deliver via rickshaw! But what finish can you use that's as eco-woke as you?

Most finishes, even the "green" ones, contain VOCs, or volatile organic compounds, which they emit as fumes or gases. Many VOCs are known to be bad for the atmosphere and for your personal health. An easy VOC test: If it smells bad, it *is* bad. Never mind the short-term side effects like headaches and nausea, VOCs have been linked to liver, kidney, and nervous system damage as well as an increase in smog and tropospheric ozone. (It's bad. You can look it up.) So our goal is to find a finish with little to no VOCs that still gets the job done.

Traditional finishes like shellacs, urethanes, and varnishes are made with alcohol and oils that are rich in VOCs, but there are safe and effective alternatives with lower VOC footprints. The more we learn about the negative effects of VOCs, the more we realize we need to find solutions. Many U.S. states are cracking down with stricter regulations, and savvy manufacturers are staying competitive by stepping up their Earth-friendly game. Here are a few of my favorite top coats from firsthand experience.

CAUTION Just because a product qualifies as "Low VOC" doesn't guarantee it's safe. It could be made of products that are harmful to you or the environment in other ways. Always read the labels and follow manufacturers' instructions for use and cleanup of their products.

Water-Based Polyurethanes

Cleaner, water-based polyurethanes have been around for a while but are becoming better and more popular. "Water poly" has a very low VOC count, can be brushed or sprayed, and cleans up easily with water. Several companies make these finishes in a variety of mixtures and thicknesses and they usually have little to no effect on the color of the wood. With patience and wet sanding in between coats, a thick, high-gloss shine that rivals the bad stuff can be achieved.

Target Coatings (targetcoatings.com) makes a fantastic line of water-based finishes including a conversion varnish that features a hybrid blend of oils and resins. I have not yet tried it but I have used their waterborne polyurethane and it performed great.

In a pinch, Varathane water polyurethane (varathanemasters.com) is usually available at the box stores and works fine. They make a "triple thick" formula that works well on very rustic wood, filling in some of the cracks and voids. It creates a plastic-like layer that you may or may not like, so try it out on a cut-off first.

TotalBoat (totalboat.com) also makes a water-based product called Halcyon that comes in clear or amber tinted formulas. It's a little thicker than most but slightly thinner than the triple thick Varathane. I find it to be "just right" for my projects. I am fond of the warm amber wood coloring, and I prefer their foil packages with resealable caps over old-fashioned cans, which reduces packaging and product waste. (Disclosure: I have a sponsorship agreement with TotalBoat so I am admittedly biased.)

Epoxy Resins

Spend any time looking at woodworking videos online and you'll see that epoxy is all the rage. It can create clear, glass-like coverings or be used with pigments to create startling, colorful effects. But is it doing more harm than good?

Epoxy resin is typically made of two separate chemical mixtures that, when combined, harden into a clear, ultra durable surface. There are certainly many types and brands out there in varying degrees of harmfulness, but more and more attention is being paid to their effects. Once cured, most epoxies are safe and many are low VOC and made with Earth-friendly ingredients.

If you need a finish that is super thick and durable, I would choose epoxy over spar urethane or other old-fashioned varnishes. Read the labels to get the right epoxy for your project; use and clean it up properly and you can still feel OK about your footprint in the morning. There are companies like EcoPoxy (ecopoxy.com), who make eco-friendly epoxies that a trustworthy peer tells me work well. ArtResin (artresin.com) also makes a "green" epoxy with rave reviews.

Working with reclaimed wood, I sometimes find myself using epoxy to fill voids and secure highly damaged parts of the wood. TotalBoat makes

Emtech by Target Coatings: They make a water-based version of all your traditional favs.

Local beeswax mixed with mineral oil; good enough to eat!

I found this cool, half-filled bottle of mineral oil at a junk shop years ago and now refill it often.

Guitars HVLP sprayed with about 10 thin coats of TotalBoat amber-tinted Halcyon.

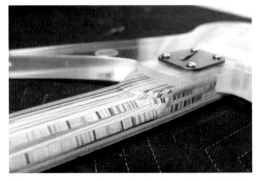

Water-based polyurethane brushed on and wet sanded between every third coat.

This reclaimed chestnut was first treated with boiled linseed oil to penetrate and bring out the natural beauty. Then, after drying for days, it was covered with a 2-part epoxy finish.

Reclaimed cumaru finished with nothing but my homemade beeswax and mineral oil polish.

many epoxies including a thin penetrating epoxy designed for this type of work that is low VOC and environmentally friendly.

Oils (the Good Kind)

Water poly is more of a top-coat that wraps a protective layer around the work. If you want something that penetrates the wood pores, you are looking for oil. Mineral oil and linseed oil are old-fashioned, tried-and-true wood finishes that are inexpensive, readily available, and not horrible for the world. Linseed oil is made from flax (it's also known as flaxseed oil) and, by itself, makes a fine finish that leaves wood darker, richer, and feeling natural. Sometimes it is mixed with turpentine and other less pleasant things, but this is not necessary. Read on for my favorite recipe.

Mineral oil is made from petroleum but highly cleaned and distilled. I buy it in my local drug store as it is also used as a laxative and relatively benign. I understand why one might not want to use anything petroleum-based but it is certainly better than many of the alternatives and is food-safe right out of the bottle. You can literally drink it!

A few years ago I melted some wax from a local beekeeper and made my own polishes. One was beeswax and linseed, the other mineral oil and wax, mixed about 50/50. I love using these and the wax adds to the protection. Telling clients you use a homemade finish is also a plus.

There are many other kinds of plant-based oil products out there like Walrus Oil (walrusoil.com), Odie's Oil (odiesoil.com) and SafeCoat (afmsafecoat.com), to name a few, but I have not used them so I cannot speak to their quality. One I have used and love is Rubio Monocoat (monocoat.us). Monocoat is a 2-part plant oil and hardener that you mix and apply with a cloth or foam brush. It is expensive but a little goes a very long way. Use it sparingly and you'll get your money's worth. Monocoat creates a fantastic, low sheen, penetrating finish that is hard to beat if you like the "au naturale" look and feel. And it smells really good.

Tung oil is another natural oil made from the nuts of the tung tree. It provides a safe and hardened finish — but beware! Many products described as tung oil are chemical concoctions that will make you dizzy, literally. While based on tung and linseed oils, Danish oil is another brew that is usually full of pretty bad stuff.

Read the Label

Perhaps the most important thing you can do is be aware of what you are buying beyond the logo and pretty pictures on the can. Many products may be packaged and named to look "green," but read the fine print. Manufacturers are required to list the ingredients. If there are words in there you have never heard, take a minute to look them up (thanks, smartphones!). You'll feel better knowing what you're using — physically and emotionally — and knowing that environmentally friendly finishes add value to your work. ◔

 To learn more about volatile organic compounds, visit epa.gov and search for VOC.

MIDI
for Makers

The 38-year-old standard for musical instrument interfacing is still a useful tool for tunes and art alike Written and photographed by Tim Deagan

A selection of the author's MIDI control surfaces.

In 1982, Hayes released the 1200 baud modem, Microsoft released MS-DOS 1.25, and the Musical Instrument Digital Interface (MIDI) was announced in *Keyboard* magazine by Robert Moog with its first implementation in the Sequential Prophet-600 synthesizer (Figure Ⓐ). While these days you're unlikely to use a 1200 baud modem or early MS-DOS, you can still use MIDI to hook up the Prophet-600 to new equipment currently being sold. Even in 2020, few manufacturers

would consider producing anything but a novelty synthesizer or drum machine without MIDI connectivity. Digital standards that can compare to this amazing 38 years of mainstream success are few and far between.

MIDI is a robust and easy-to-understand protocol. Simple messages are sent and received without handshakes or acknowledgments. While slow by today's standards, for the purpose of sending musical information it still functions

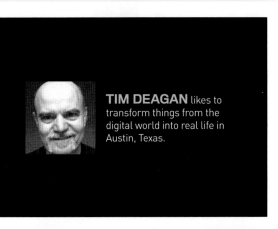

TIM DEAGAN likes to transform things from the digital world into real life in Austin, Texas.

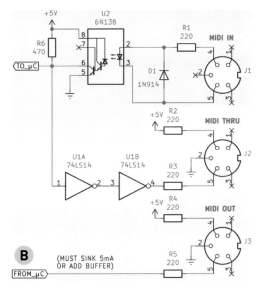

A

B (MUST SINK 5mA OR ADD BUFFER)

FROM_μC

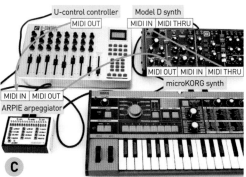

C

effectively for all but the most exotic applications. Given the speeds of modern microcontrollers, building MIDI devices is easy, cheap, and extremely satisfying. A quick perusal of the Tindie marketplace for MIDI devices underscores the creative ideas that continue to explode in the maker community.

Makers can also gain access to the universe of MIDI control surfaces. Grid controllers, slider and knob boxes, drum machines, and even wind controllers that output MIDI can become parts of projects that may have nothing to do with music. Capturing and repurposing their messages is a simple matter of programming. To understand what this would take, let's look closer at the standard itself.

How It Works

MIDI combines a messaging standard, a digital protocol for sending those messages, and a physical scheme for how to wire systems together. This article will focus on the "classic" MIDI standard that uses 5-pin DIN plugs, and sometimes ⅛" (3.5mm) tip-ring-sleeve (TRS) plugs, with a current loop connection scheme. MIDI has also become popular over USB, but that's an approach for another article.

Starting with the physical side of things, MIDI uses unidirectional *current loop* networking. This means that sending 1s and 0s is done with changes in current rather than voltage. The MIDI standard limits the current loop signals to 5mA and 5V, so realistically runs of MIDI cable can easily be 20 feet, and can more than double that with high quality cables. MIDI also uses a common current loop technique of opto-isolating the receiving system to prevent ground loops and voltage spikes (Figure **B**). MIDI circuits are defined as MIDI IN, MIDI OUT and MIDI THRU. OUT and THRU differ only in that THRU is wired to IN and sends along whatever is received (Figure **C**). This ability to "chain" MIDI devices is one of the great features MIDI users enjoy.

While wiring MIDI circuits is straightforward, multiple vendors produce inexpensive MIDI shields for Arduino (usable as breakout boards for other microcontrollers). These boards typically provide standard serial I/O connections for the host controller and include the opto-isolator and

DIN plugs (Figure **D**).

Moving up from the physical, MIDI operates at 31250 baud. This value has often frustrated users since it is, to say the least, uncommon to find it implemented outside of MIDI. In actuality, its origins make perfect sense. The first MIDI implementations commonly used the outstanding Intel 8051 microcontroller. The 8051, at least in 1982, used an external XTAL oscillator with 12MHz as the fastest supported value. Internally, it divided that signal by 12 to derive a 1MHZ timer clock. If you divide 1MHz by 32 (2^5), you get 31,250!

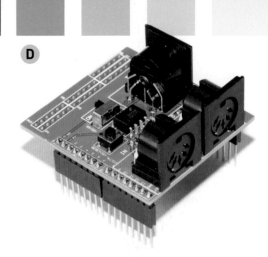

D

E	**STATUS**								**DATA 1 (if needed)**								**DATA 2 (if needed)**							
	1	t	t	t	n	n	n	n	0	x	x	x	x	x	x	x	0	x	x	x	x	x	x	x
STATUS		Type of Message			Channel # 1-16 (Values 0-15)				**DATA**	Data Value 1 (0-127)							**DATA**	Data Value 2 (0-127)						

HEX						
0x8n	1	0	0	0	Note OFF	
0x9n	1	0	0	1	Note ON	
0xAn	1	0	1	0	Polyphonic Aftertouch	
0xBn	1	0	1	1	Control Change (CC)	CHANNEL VOICE MESSAGES
0xCn	1	1	0	0	Program Change	(CC Controllers 120-127 reserved for CHANNEL MODE MESSAGES)
0xDn	1	1	0	1	Channel Aftertouch	
0xEn	1	1	1	0	Pitch Wheel	
0xFn	1	1	1	1	SYSTEM MESSAGE	Common, Real-Time, Exclusive

MIDI Messages

MIDI encodes its messages into a *status byte* which describes the type of message, followed by up to two *data bytes* (Figure **E**). Each data frame has a start bit (value 0,) and a stop bit (value 1,) for a total length of ten bits per byte sent. Status bytes have a 1 as the most significant bit (MSB) and data bytes have a 0 as the MSB.

Messages can be *channel messages* which are targeted at one of the 16 channels MIDI supports (types: channel voice, channel mode) or *system messages* received by all devices (types: system common, system real-time, and system exclusive.) Channel messages encode the message type onto the three bits following the MSB and the target channel onto the 4 Least Significant Bits (LSB) of the status byte. Data bytes encode their payload onto the 7 bits

following the MSB.

A full description of all the different MIDI messages is beyond the scope of this article, but it's worth looking at a couple of common messages (Figure **F**). Let's say we want to send a message to a synthesizer on channel 4 to turn on the note C3. (Having 7 bits to encode note number means that MIDI has a range of 127 notes, from C-1 to G9.) We'd send a status byte with the channel and message type, a data byte with the note number and a second data byte with the velocity (the force with which the note is played). If we wanted to change the value of a parameter, such as the modulation wheel, then we'd send a status byte, a data byte with the controller number, and a data byte with the controller value we want to set.

Example MIDI Messages	(Values in Hex)		
	Status Byte	Data Byte 1	Data Byte 2
Note ON, CH 4, C3, Max Velocity	0x93	0xB0	0x7F
Note OFF, CH 4, C3, Max Velocity	0x83	0xB0	0x7F
Program Change, CH4, Pgm 10 (dec)	0xC3	0x0A	-
Control Change, Pan, CH 4, Hard Left	0xB3	0x0A	0x00
System Realtime, START	0xFA	-	-
System Realtime, STOP	0xFC	-	-

```
MIDI
#include <MIDI.h>
MIDI_CREATE_DEFAULT_INSTANCE();
/*······································*/
void setup(){
    // SETUP MIDI
    if (MIDIChan < 1){
        MIDI.begin(MIDI_CHANNEL_OMNI);// Launch MIDI,listen to all channels
    } else {
        MIDI.begin(MIDIChan);
    }
}

void loop(){
    if (MIDI.read())             // If we have received a message
    {
        switch(MIDI.getType())      // Get the Message Type
        {
            int rcvNote;
            case midi::NoteOn:      // NoteOn message received
                rcvNote = MIDI.getData1(); // get MIDI Note #
                // Do some fun stuff!
                break;
            case midi::NoteOff:     // NoteOff message received
                rcvNote = MIDI.getData1(); // get MIDI Note #
                // Do some more fun stuff!
                break;
            case midi::ProgramChange:
                break;
            case midi::ControlChange:
                break;
            case midi::AfterTouchPoly:
                break;
            case midi::AfterTouchChannel:
                break;
            case midi::PitchBend:
                break;
            default:
                break;
        }
    }
}
```

G

MIDI PROJECT:
Building BooshBeats
An Arduino sequencer using Novation Launchpad Pro

Who doesn't love a grid of pads with RGB LEDs? When I needed to create an 8-channel × 8-step sequencer for my mega-flame effect BooshBeats (makezine.com/go/boosh-beats), I used MIDI to talk back and forth with a Novation Launchpad Pro. Button presses came as MIDI messages in, and I could control the colors of the pads with MIDI messages out. The Launchpad grid visually displays the pattern, and lets me edit existing and create new patterns, even while playing! I use the top, bottom, and left buttons to start, stop, retrieve, and store patterns. The right side buttons are used as displays to show the sequencer walking through the steps.

Building Projects

Much like deciding to use a MIDI shield instead of wiring your own interface, using MIDI libraries like the one from FortySevenEffects (github.com/FortySevenEffects/arduino_midi_library) makes it extremely easy to handle all the technical details of sending and receiving messages. Once messages are received, a case structure or chain of **if** statements selects the desired logic to implement. Sending messages requires only a single command (Figure G).

Advanced MIDI messaging such as MIDI timing/clock and System Exclusive (sysex) are all supported in the FortySevenEffects library. But fantastic MIDI projects may only need to use a couple of messages like **NoteOn** and **NoteOff** or a small set of **ControlChange** messages. Just ignore any messages that you aren't interested in.

Any sensor you can attach to a microcontroller can become a substitute for the keys or controls of a synth or the pads of a drum machine. Simple sketches can transpose notes, create arpeggiations, reroute messages, or any other operation you can perform on MIDI's messages.

In 2019, MIDI 2.0 was announced, assuring the future of the standard. The new standard will be backward compatible and support existing MIDI devices, assuring the viability of maker MIDI projects for a long time to come. ◗

 Learn more about MIDI
- midi.org/specifications-old/item/table-1-summary-of-midi-message
- learn.sparkfun.com/tutorials/midi-tutorial/all

ROLAND LEF 12-I FLATBED UV PRINTER

$17,995 rolanddga.com/products/printers/versauv-lef-12i-flatbed-printer

This printer can print on almost anything. I know that sounds a bit odd, but that's the incredible power of a UV printer. The Roland VersaUV LEF 12-i uses microscopic jets to shoot UV-curable ink onto a surface up to a few millimeters away and blasts it with UV light to cure it, meaning that not only will it print on things like glass, metal, wood, fabric, and plastic, it can print on uneven and textured surfaces as well. Want to put your name on a golf ball? This thing can do it.

To take things to another level of creative potential, this printer can also layer a clear material up to create texture of its own. Let's say, for example, you want a logo with raised lines that you can feel with your finger, or even the texture of wood, this printer can do that too.

Of course, all those features come with a hefty price tag, which means a UV flatbed printer may not be a great fit in your typical maker's garage. These typically reside in print shops, and I've seen them in use in makerspaces as well. —*Caleb Kraft*

ANDOER 2400W SOFTBOX LIGHTING KIT

$110 amazon.com/gp/product/B07QDH51VQ

When it comes to shooting video or photos in your home, it can be hard to shell out cash for a decent light setup. This is especially true if your cinematography and photography are relegated to hobby status. However, anyone who tinkers with celluloid (or the digital analog of that wondrous material) knows that light can make or break your results.

The Andoer model "F" softbox lighting kit comes with everything you need to have a solid 3-point light setup in a portable package. The system uses CFL bulbs which can be easily replaced as they die, and has two separate switches per light allowing for rudimentary brightness control. The stands that come with it are a bit flimsy, but they do the job and can actually be a great addition for things like holding tiny action cameras for overhead shots. The softbox diffusers are quick to assemble and give a nice, even soft light, which is pretty much all you can ask for. —*Caleb Kraft*

CERAMBOT PRO

$699 cerambot.com

While 3D printing is pretty amazing, and machines have come very far in terms of quality, there's still a moment that I find to be a little bit of a letdown with each print — when I pick it up and it just feels like a hunk of plastic in my hand. This is somewhat unavoidable on account of the fact that it *is* plastic. The Cerambot is configured for extruding ceramic, which feels fantastic when you pick it up, and is advertised at roughly half the price of other ceramic printers I've seen, so I picked one up.

Mine was a kit, and at an even more reduced price than the one listed on this review, but since that price was for Kickstarter, and no longer available, I'm not listing it. This kit is somewhat of a bare-bones experience. The instructions are minimal and the parts are minimal as well. If you have a solid understanding of how a 3D printer goes together and works, you'll figure it out, but I wouldn't recommend it for someone's first experience. The prints that come off of this, and any ceramic printer for that matter, are as much art as science, requiring intensive trial and error. You're not going to get the same detail as you would from a normal 3D printer, but boy do those ceramic prints feel *good*. —*Caleb Kraft*

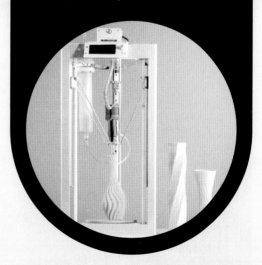

ROLLING ROBOTS

DJI ROBOMASTER S1 **$549** www.dji.com/robomaster-s1
SPHERO RVR **$249** sphero.com/rvr

Young minds looking to engage with a robotics platform can try to build their own from scratch, or jump into a full kit or premade device and move quickly to programming. Sphero's RVR and DJI's RoboMaster S1 are two new devices that allow for the latter, with a fair amount of shared aspects and some key differences.

The S1 is a larger rig than the RVR, although both could be considered substantial. It includes a gimbaled camera and beam/pellet shooter sit on top, and four beam/impact sensors on its base. Its four omnidirectional Mecanum wheels move it in any direction without turning. The aggressive styling and shoot-em-up tools are derived from DJI's RoboMaster competition series.

RVR is a less-outfitted (and much cheaper) machine — essentially a tank-tread bare platform that you can race around or dig into for robotics tinkering.

Both systems provide quick access to Scratch-style block programming, and their apps include tutorials for this. You can take programming further with more advanced Python code, but for the RVR you'll need to supply an external board — it supports Raspberry Pi, Arduino, and Micro:Bit, and comes with a board-mountable roof that you can swap onto it.

Where each platform pulls away from other programming-focused toys is their accessible electronics. Both offer UART and USB interfaces. The S1 has 6 PWM ports and an SBus port built in, while those can be added to the RVR through an external board like those mentioned above. This allows a user to add external hardware and, depending on your programming savvy, set it to work within the system. Add-on AI camera kits from SparkFun are already available for the RVR, for instance.

Which is best for you? That depends. Do you want a more complete, refined system that's largely designed for competition, or a cheaper setup designed for you to hack away on? Both give you a lot of control disguised as fun.

—Mike Senese

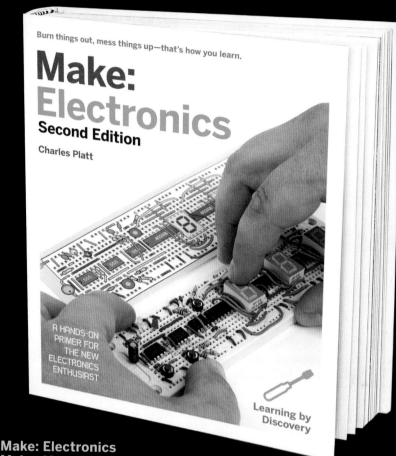

AUTOMATA MAGAZINE

$60 automatamagazine.com

I never would have imagined that there would be a magazine focusing purely on automata, or that it would be this delightful.

The first two issues of this digital-only magazine are available as free downloads. Check them out and you'll find an eclectic mix of mechanical wonders, from simple beginner projects up to the massive 40-foot-tall walking mechanical elephant of Les Machines de l'Île in Nantes, France.

Most of the articles are stories of artists, and visual explorations of their work, but you will find some enlightening examples of how they pulled off certain movements. There are also some step-by-step tutorials to get you going, and tips to improve your craft.

—Caleb Kraft

A DUO OF RAMEN BOOKS

Magic Ramen: The Story of Momofuku Ando (2019, Simon & Schuster) tells the story behind the iconic (instant) food. Andrea Wang and illustrator Kana Urbanowicz detail Ando's methodical efforts in his backyard shed to create an inexpensive, accessible version of a food that had become a luxury in post-war Japan. Too crumbly, too sticky, too lumpy, too tasteless ... Ando tried it all before bringing his flash-frying flavor-packet formula to market in 1958 hoping that "peace follows a full stomach."

For the make-at-home version, chef Hugh Amano and illustrator Sarah Becan's colorful, instructive **Let's Make Ramen** (2019, Ten Speed Press) shows you step-by-step how to up your game from Cup Noodles. Like a choose-your-own-adventure graphic novel, they activate their "ramen twin" powers to outline different styles and preparation tips that show you how to make your ultimate bowl alongside cultural insights about all things ramen.

—Jennifer Blakeslee

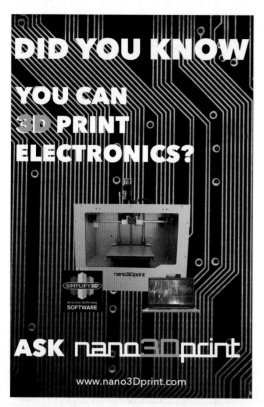

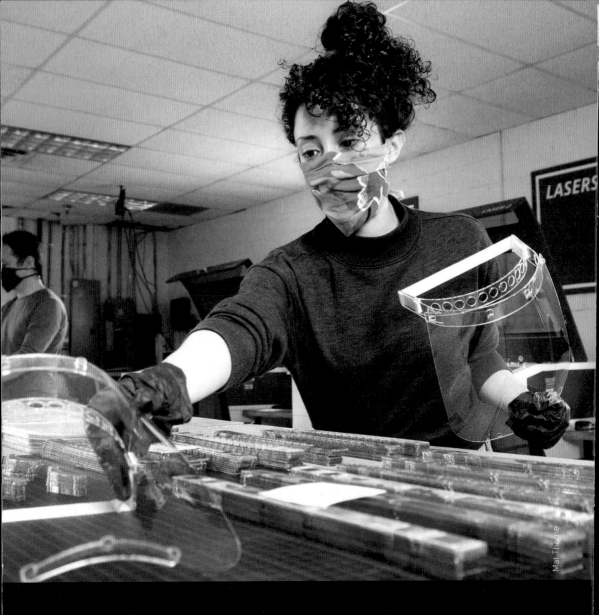

To the health workers, the PPE makers,
and everyone else in this fight, thank you.
Your hard work makes a difference. We're proud
and amazed to be part of your community.

YOU ARE TRUE HEROES!

—Make: